MARINE BIOGEOCHEMICAL CYCLES

THE OCEANOGRAPHY COURSE TEAM

Authors
Evelyn Brown (*Waves, Tides, etc.*)
Angela Colling (*Ocean Circulation*; *Seawater* (*2nd edn*); *Waves, Tides, etc.* (*2nd edn*); *Case Studies*)
Rachael James (*Marine Biogeochemical Cycles*)
Dave Park (*Waves, Tides, etc.*)
John Phillips (*Case Studies*)
Dave Rothery (*Ocean Basins*)
John Wright (*Ocean Basins*; *Seawater*; *Waves, Tides, etc.* (*2nd edn*); *Marine Biogeochemical Cycles*; *Case Studies*)

Other members
Mark Brandon
Kevin Burton

Course Manager
Jennie Neve Bellamy

Designer
Jane Sheppard

Graphic Artists
Roger Courthold
Steve Best

Cartographer
Ray Munns

Editor
Gerry Bearman

This Volume forms part of an Open University course, *S330 Oceanography*. For general availability of all the Volumes in this series, please contact your regular supplier, or in case of difficulty the appropriate Butterworth–Heinemann office.

Further information on Open University courses may be obtained from: The Student Registration and Enquiry Service, The Open University, P.O. Box 625, Walton Hall, Milton Keynes MK7 6YG, UK or from the Open University website: http://www.open.ac.uk

Cover illustration: Satellite photograph showing distribution of phytoplankton pigments in the North Atlantic off the US coast in the region of the Gulf Stream and the Labrador Current. (*NASA, and O. Brown and R. Evans, University of Miami.*)

MARINE BIOGEOCHEMICAL CYCLES

PREPARED BY RACHAEL JAMES FOR THE COURSE TEAM

ELSEVIER
BUTTERWORTH
HEINEMANN

in association with

The Open University

THE OPEN UNIVERSITY, WALTON HALL,
MILTON KEYNES, MK7 6AA, ENGLAND

Elsevier Butterworth–Heinemann
Linacre House, Jordan Hill, Oxford OX2 8DP
A division of Reed Educational and Professional Publishing Ltd

AMSTERDAM BOSTON HEILDELBERG LONDON NEW YORK
OXFORD PARIS SYDNEY SAN DIEGO SAN FRANCISCO
SINGAPORE

British Library Cataloguing in Publication Data
A catalogue record for this book is available from the British Library

ISBN 0 7506 6793 1

Library of Congress Cataloguing in Publication Data
A catalogue record for this book is available from the Library of Congress

Jointly published by the Open University, Walton Hall, Milton Keynes
MK7 6AA and Elsevier Butterworth–Heinemann

Edited, typeset, illustrated and designed by The Open University
Printed in Singapore by Kyodo under the supervision of MRM
Graphics Ltd., UK

s330v5i2.1

CONTENTS

ABOUT THIS VOLUME

This is one of a Series of Volumes on Oceanography. It is designed so that it can be read on its own, like any other textbook, or studied as part of S330 *Oceanography*, a third level course for Open University students. The science of oceanography as a whole is multidisciplinary. However, different aspects fall naturally within the scope of one or other of the major 'traditional' disciplines. Thus, you will get the most out of this Volume if you have some previous experience of studying geology and a certain amount of chemistry. Other Volumes in this Series lie variously within the fields of geology, biology, physics or chemistry.

Chapter 1 begins by explaining why the study of the occurrence, distribution and cycling of chemical elements within the ocean is of contemporary interest, and goes on to examine one of the most important components of marine biogeochemical cycles, the sea-floor sediments.

Chapter 2 reviews the steady-state ocean concept and explains how different constituents of seawater can be classified according to their involvement in the biological particle cycle (formation, destruction and regeneration of organic material), which is also a major control on residence times in the oceans. A simple two-box model of ocean cycling is introduced and the role of dissolved gases is described, with special reference to oxygen.

Chapter 3 discusses how water chemistry controls the extent to which the remains of calcareous and siliceous organisms dissolve when they sink to the sea-bed and form part of the sediments. This Chapter also explores how sediments that come from continental areas may be transported to the deep sea.

Chapter 4 explains how information about the history of the oceans can be obtained from sediments recovered in deep-sea drill cores. This is an increasingly important aspect of oceanography because it tells us how the oceans have affected Earth's climate in the past, which helps us to predict how climate might change in the future.

Chapter 5 provides a review of biogeochemical processes occurring within marine sediments. The uppermost layers of sediment are disturbed and disrupted by the activities of marine organisms (the benthos) and by the effects of bottom currents which can resuspend and redistribute sedimentary material. In addition, there are important chemical exchanges between seawater and sediment, as a result of reactions occurring within the main body of the sediment pile.

Finally, you will find questions designed to help you develop arguments and/or test your understanding as you read, with answers provided at the back of this Volume. Important technical terms are printed in **bold** type where they are first introduced or defined.

ABOUT THIS SERIES

The Series as a whole (see back cover for other titles) provides an interdisciplinary introduction to marine science. There is also a companion Volume to the Series by C. M. Lalli and T. R. Parsons entitled *Biological Oceanography — An Introduction* which has a similar style and format and is set reading for Open University students.

CHAPTER 1   INTRODUCTION

While the Earth receives a continuous supply of energy from the Sun, the Earth's store of materials (excluding small amounts of extraterrestrial matter) is fixed and finite. This means that, for life to continue, Nature must recycle biologically important elements. During cycling, the form of these elements continually changes as one chemical compound is transformed into another while passing through the Earth's chemical reservoirs: the atmosphere, land and ocean.

The study of the occurrence, distribution and cycling of chemical elements within the ocean is therefore central to understanding the Earth's life support system. Today, this is more important than ever before because the chemical cycles of many elements — particularly carbon, but also others such as copper, lead, cadmium, mercury, nickel and silver — are being grossly perturbed by human activity. In addition, there are many new substances in the marine environment that were not even there a century ago. These include pesticides and other organic chemicals, as well as transuranic and other artificial nuclides from nuclear weapons testing and low-level waste discharges.

We also need to clarify how marine biogeochemical cycles have changed over time. This is because we know that the Earth's capacity for supporting life has changed; for example, the rise of animals at around 600 Ma ago has been linked with large increases in levels of atmospheric oxygen at that time. Fortunately, analysis of the physical, chemical and biotic properties of marine sediments, which accumulate layer upon layer, year after year, allows us to build up records of various aspects of the environment of the ocean in which the sediment particles temporarily resided (Figure 1.1). Let's begin, then, by looking at the sediments.

1.1   THE NATURE OF DEEP-SEA SEDIMENTS

When HMS *Challenger* (Figure 1.2) returned to England on 24 May 1876 laden with specimens, records and measurements after an epic three years and nine months voyage of oceanic exploration, the era of systematic oceanography had begun. The member of the scientific party who did most to ensure world recognition of *Challenger*'s scientific achievements was John Murray, a Canadian-born Scot, who owed his place on the ship to chance, when a member of the original team was obliged to drop out at

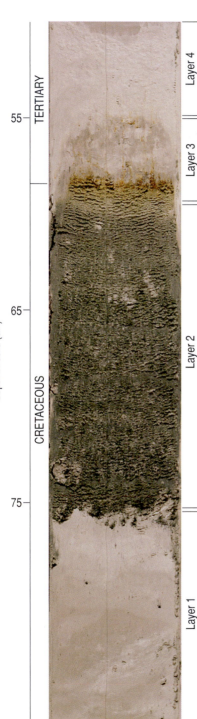

Figure 1.1   Cretaceous–Tertiary boundary recorded in a deep-sea sediment core recovered from the subtropical northwestern Atlantic. The lowermost layer (Layer 1) comprises soft, microfossil-rich sediment (foraminiferal ooze). This is overlain by a layer of green glassy globules, called tektites, as well as mineral grains and rock debris apparently derived from a meteorite impact crater (Layer 2). The thin bed on top of this layer (Layer 3) is devoid of almost all signs of life, and contains the remains of the meteorite. Repopulation of the oceans in the Early Tertiary is recorded by the reappearance of microfossils (Layer 4).

short notice. Murray's account of the samples collected from the floor of the oceans (Figure 1.3) provided the starting point for all subsequent investigations into deep-sea sediments.

Figure 1.2   HMS *Challenger*, 1872. She was a steam-assisted wooden corvette of 2306 tonnes.

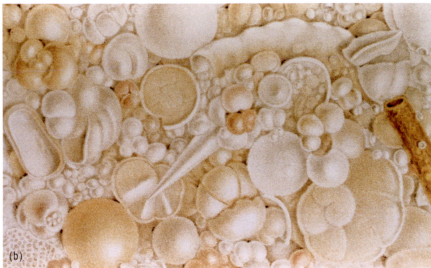

(b)   Washed foraminiferal ooze, mainly *Globigerina* spp. Other shells include a pteropod (central cone-shaped shell) and an ostracod (planktonic crustacean, extreme left centre). Width of field *c*. 3.5 mm. Sample from 1900 fathoms (*c*. 3500 m).

Figure 1.3   Drawings of planktonic remains in sediments collected from the North Atlantic during *Challenger*'s voyage.
(a)   Drawings of radiolarians of the genus *Hexastylus*.

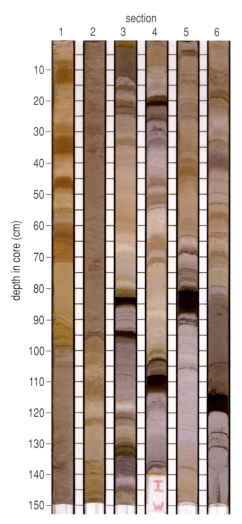

section

1 2 3 4 5 6

depth in core (cm)

Figure 1.4   Sediment core recovered from between 27 and 36 m below the sea-floor in the eastern Mediterranean (35° 46.8′ N 18° 56.9′ E). The core has been divided into six sections; the top of the core is at upper left and the bottom is at bottom right. The water depth at this site is ~3700 m. Sediments consist of nanofossil ooze, which is grey to pale brown in colour. The black layers, between 1 and 7 cm thick, are sapropels. These are sediments that are unusually rich in organic carbon (up to 30 wt %), and are apparently deposited under anoxic conditions (see Chapter 5).

At first sight, deep-sea sediments are little more than soft muds of various hues, from white to grey to reddish-brown (Figure 1.4). Two main kinds of deep-sea sediments can be recognized.

Terrigenous sediments are formed by weathering and erosion of land areas, and are transported to the oceans by rivers, glaciers and wind. They comprise gravels and sands, and smaller-sized particles including silts and clays. Biogenic (or biogenous) sediments are made up of the microscopic remains of those predominantly **planktonic** marine organisms that secrete skeletons (or tests) of calcium carbonate or silica. At this point, we must emphasize that there is no such thing as a 'pure' terrigenous or biogenic sediment. For example, terrigenous dust is widely dispersed by winds and currents, so biogenic sediments always contain material of non-biogenic origin, and conversely, terrigenous sediments are seldom without a biogenic component, however small.

There is a convenient 'rule of thumb' for broadly classifying deep-sea sediments, known as the '30% rule'. If the sediment contains more than 30% biogenic components, it is called a calcareous or siliceous ooze (depending on which biogenic component is dominant); if it contains less than 30%, it is a clay.

1.1.1   BIOGENIC SEDIMENTS

Viewed through a microscope, biogenic sediments are seen to consist of a wonderful variety of delicate and intricate structures, mostly of the skeletal remains of marine phytoplankton and zooplankton. As the lifetime of most of these planktonic organisms is only about a week or two, there is a slow but continuous 'rain' of their remains down through the water column that builds up successive layers of sediment. As you will see shortly, the presence of the remains of any one particular type of planktonic organism depends on a number of local factors such as water chemistry at depth, and the nutrient concentration and the extent of primary production in the surface ocean waters. Because of this, the presence of these microfossils in ancient deep-sea sediments can be used to determine what the water depths and surface productivity were like during the geological past (see Chapter 4).

Carbonate sediments are composed principally of the skeletal remains of **coccolithophores**, **foraminiferans** and **pteropods**. Siliceous remains come mostly from **diatoms** and **radiolarians**. The hard parts of these organisms vary a good deal in size, shape and chemical stability, all factors that determine how well they are preserved in deep-sea sediments.

Coccoliths are minute plates of **calcite** (the more common form of calcium carbonate, $CaCO_3$), usually less than 10 µm in size (µm = micrometre (micron) = 10^{-6} m), with which the phytoplanktonic (algal) coccolithophores envelop themselves to form coccospheres (Figure 1.5). When the algal cell dies, the coccosphere disintegrates easily, releasing the individual plates — the coccoliths — into suspension. Thus, it is the coccoliths rather than the whole coccospheres that are preserved in sediments. Plates are also shed from the coccospheres as the algae grow, so coccolithophores contribute to deep-sea calcareous sediments before the death of the organism, as well as afterwards. Each coccolith has an organic membranous covering which inhibits dissolution of the calcite and enhances preservation. The white cliffs of Dover are formed of chalk that is mostly composed of coccoliths.

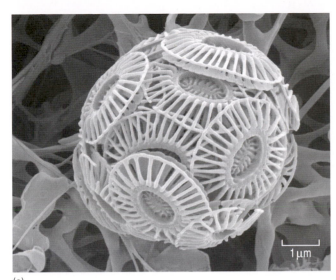

(a)

(b)

Figure 1.5   (a) Coccosphere of the coccolithophore *Emiliania huxleyi.* (b) Satellite image of a coccolithophore (*Emiliania huxleyi*) bloom in the English Channel off the south coast of Cornwall, 24 July 1999.

Because of their exceptionally small size, coccoliths are referred to as **nanofossils** (from the Greek word 'nanos', meaning dwarf) and calcareous sediments particularly rich in coccoliths are known as **nanofossil oozes** (or simply as coccolith oozes or nano-oozes).

Foraminiferans produce calcite exoskeletons, or tests, in the size range 30 μm to 1 mm (Figure 1.6). Most zooplanktonic foraminiferans live in the top 1000 m of the water column. Surface and near-surface species tend to have a more spiny shape, which increases the ratio of surface area to volume and thereby aids a floating mode of life, as well as deterring predators. However, an abundance of spines also increases the surface area of the test, making it more prone to dissolution after death. Deeper-water species, which usually lack spines, are more commonly preserved. Because of their relatively large size, abundant foraminiferans give a sandy texture to the sediments in which they occur. There are also both shallow-water and deep-water forms of bottom-living or **benthic** foraminiferans, and in some regions these can be abundant.

Figure 1.6   Planktonic foraminiferans. (a) *Globigerina bulloides.* This species is mainly found in subpolar waters, and also in regions of upwelling. The calcite skeleton consists of four spherical chambers, and has an open, arched aperture. (b) *Globorotalia menardii.* This species is mainly found in subtropical waters. The calcite skeleton consists of five or six wedge-shaped chambers, and has a prominent outer crust known as a 'keel'.

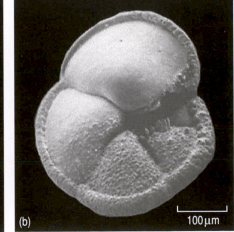

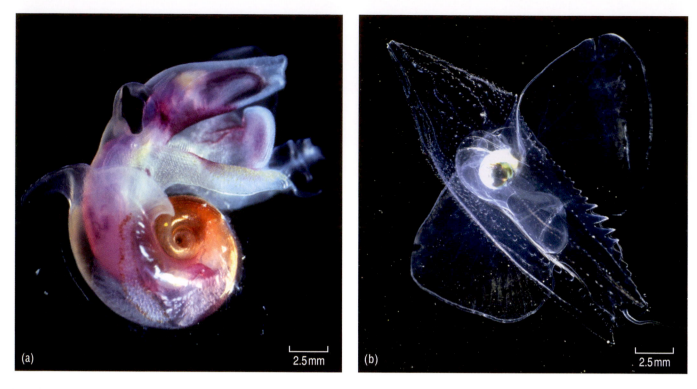

(a)

2.5mm

(b)

2.5mm

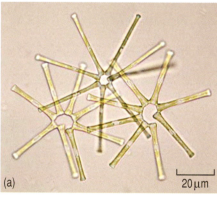

(a)

20μm

Figure 1.7 Living pteropod molluscs. (a) *Candida atlanta*. This marine snail has a transparent shell, and paired muscular swimming wings protrude from its body. (b) Venus slipper (*Cymbula* sp.). The Venus slipper lacks a true shell; instead, it has an internal skeletal structure. Like all pteropods, it swims by means of the paired wings that can clearly be seen on either side of its body.

Pteropods are planktonic molluscs that are up to a centimetre long (Figure 1.7). At the present day, most species are restricted to tropical and subtropical oceanic areas. Most have thin shells which are composed of **aragonite** (a variety of calcium carbonate, $CaCO_3$, that is more soluble than calcite), so they are more easily dissolved than calcitic plankton shells and are not found in ocean sediments where water depths exceed about 2–3 km.

Diatoms are algae ranging in size from a few μm to around 200 μm (Figure 1.8). They are single-celled (but may form chains or colonies) and secrete shells (called **frustules**) of amorphous hydrated silica (opaline silica, or opal), the formula of which is commonly written as $SiO_2 n H_2O$. Both planktonic and shallow-water benthic diatoms occur, but the planktonic species have thinner tests and are more prone to dissolution.

QUESTION 1.1 Why are benthic diatoms restricted to sediments lying in shallow water?

Radiolarians are quite large zooplanktonic organisms (Figure 1.9), usually between 50 μm and 300 μm or more in size. They also have skeletons of silica and are the dominant biogenic component of siliceous sediments found at low latitudes. As with the foraminiferans, both surface and deeper-water species occur.

QUESTION 1.2 (a) By analogy with foraminiferans, how would you expect the shape and preservation potential of surface forms of radiolarians to differ from those of deeper-water forms? (b) In a typical sample of siliceous sediment, would you expect to find proportionally more radiolarians, or more diatoms, assuming that both are present in overlying seawater?

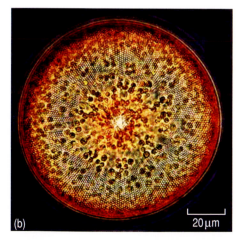

(b)

20μm

Figure 1.8 (a) A colony of diatoms, *Asterionella*. (b) A centric diatom, *Coscinodiscus*. This species is large, and can just be seen with the naked eye.

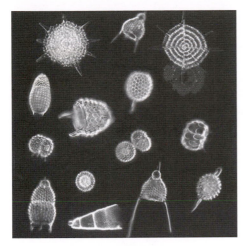

Figure 1.9 Examples of skeletons of radiolarians from a washed siliceous ooze. Individual tests *c.* 100 μm across.

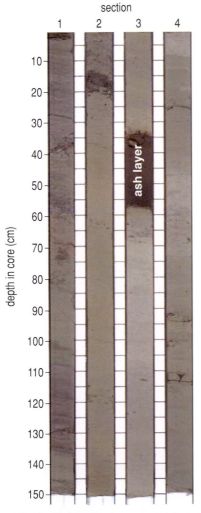

Figure 1.10 Sediment core recovered from the Norway Basin (66° 56.5′ N, 6° 27.0′ W) at a water depth of 2788 m. The core has been divided into four sections; the third contains a ~20 cm thick ash layer that was derived from a major volcanic eruption on Iceland ~3.95 Ma ago.

It is important always to keep in mind that the distribution of biogenic sediments is determined partly by the extent of biological productivity of the plankton in surface waters; and partly by the extent to which the skeletal remains are dissolved in the water column and (particularly) at the sea-bed — and that in turn depends upon the chemical properties of seawater (see Chapter 3).

1.1.2 TERRIGENOUS SEDIMENTS

Nearly all terrigenous sediments deposited from the deep sea (**pelagic** terrigenous sediments), beyond the influence of processes along continental margins, are composed of material of the smallest grain sizes. There are two ways in which coarser-grained material such as sands, gravels and boulders can reach the deep-sea (abyssal) environment. One is via turbidity currents and other gravity flows (see Chapter 3). The other way is through ice-rafting, that is, glacial material being shed by icebergs drifting into the open ocean and melting. Ice-rafted boulders, pebbles and sand may be found among deep-sea sediments up to several hundred kilometres from the glaciers of ice-shelves from which they originally 'calved'.

The wind is obviously a powerful means of transporting fine material directly to the open oceans. Indeed, it is important to emphasize that clays deposited in the open oceans are predominantly of wind-blown (aeolian) origin and mostly less than about 20 μm in size. The regions most likely to generate wind-blown dusts are the low-latitude belts influenced by persistent Trade Winds and low rainfall. The total amount of wind-blown dust delivered to the oceans annually is of the order of 10^8 tonnes. This is very small compared with the sediment load supplied by rivers, which is about 1.5×10^{10} tonnes per year. Some river-borne sediments do reach the abyssal plains (see Chapter 3) but nearly all of it is deposited along continental margins. That includes clay minerals which are largely removed from suspension by **flocculation** in river mouths and estuaries. So the proportion of river-borne terrigenous sediment in deep-sea clays is usually small relative to the aeolian contribution.

Wind-blown dust is continually being supplied to the oceans, and it consists mostly of clay minerals, along with a small proportion (up to about 10%) of minute quartz particles and organic remains. More intermittent and much more spectacular is the supply of volcanogenic material to the pelagic environment, though the proportion of such material in deep-sea sediments is small on a global scale. Major eruptions, such as that of Krakatoa in 1886, can eject large quantities of volcanic ash and dust up to heights of 15 to 50 km, where the smallest particles, 1 μm or less, can remain suspended for many months. During that time, they can be carried several times around the world by high altitude winds, causing unusual weather conditions and spectacular sunsets. Material between about 1 μm and 20 μm in size will rarely be projected to heights above 10 km, and is deposited in a matter of days or weeks and within several hundred to a few thousand kilometres from the eruption. The result is distinctive layers of volcanic ash, which can be useful in the correlation and hence dating of pelagic sediment sequences from widely separated locations (Figure 1.10).

There are a growing number of wind-borne anthropogenic contributions to sea-bed sediments, both shallow and deep. Some of these are benign, others less so. They include dust from power stations and cement

3μm

Figure 1.11 Scanning electron micrograph of a fly-ash particle from a power station embedded in biogenic debris and clay minerals, collected in a sediment trap on the floor of the Sargasso Sea.

works (Figure 1.11); particles of more persistent plastics; PCBs (polychlorinated biphenyls from the plastics and electrical industries); lead compounds (mainly from motor vehicles); radionuclides; and various products from waste incineration at sea. Production of these materials has been limited to the past hundred years or less, so their appearance in sediments can be used as a time-marker and in some cases also as a tracer for the movement of material through the oceans, on its way to the sea-bed.

Clay minerals are usually the products of weathering of terrestrial rocks. They are rich in silica and aluminium, and they occur as thin flakes, generally less than 2 μm across. There are four main types of clay minerals in deep-sea sediments — kaolinite, chlorite, illite and montmorillonite — each of which is formed in a different weathering environment. Their relative proportions in a pelagic clay vary according to the prevailing climatic and geological conditions in the source region and along transport pathways, and according to the mixing processes that occur in the oceans.

Kaolinite is formed by the extreme chemical weathering of silicate minerals, especially feldspars, and is most abundant in low latitudes.

Chlorite occurs in both igneous and metamorphic rocks of the continents, but is destroyed by the chemical weathering that predominates in low latitudes. Chlorite is therefore most common in pelagic clays at high latitude, where physical weathering predominates and chlorite is released to the ocean in an unaltered state.

Illite is the most widespread clay mineral, but is more abundant in the Northern Hemisphere, where it may contribute up to 70% of the clay minerals in a sediment. Illites form under a variety of conditions and are not characteristic of any particular latitude belt; so whether they dominate or not depends on the degree of dilution with other clay minerals.

Montmorillonite is an alteration product of volcanogenic material. Much of the montmorillonite in deep-sea sediments is produced by the 'weathering' of volcanic ash actually on the sea-floor. Thus, strictly speaking, it is not always terrigenous material, because some of it is formed *in situ* (see Section 5.1.1).

1.2   THE DISTRIBUTION OF DEEP-SEA SEDIMENTS

The distribution of deep-sea sediments is summarized in Figure 1.12 and Table 1.1, and Figure 1.13 is a map of the major physiographic features of the ocean basins.

Table 1.1   Percentage of deep ocean floor covered by pelagic sediments.

Sediments	Atlantic	Pacific	Indian	World
Calcareous ooze	65.1	36.2	54.3	47.1
Pteropod ooze	2.4	0.1	–	0.6
Diatom ooze	6.7	10.1	19.9	11.6
Radiolarian ooze	–	4.6	0.5	2.6
Clays	25.8	49.0	25.3	38.1
Relative size of ocean (% of total)	23.0	53.4	23.6	100.0

QUESTION 1.3   Examine Figures 1.12 and 1.13. Is there any relationship between the distribution of calcareous sediments, siliceous sediments, and clays, and the main physiographic features? Where is most of the terrigenous sediment found?

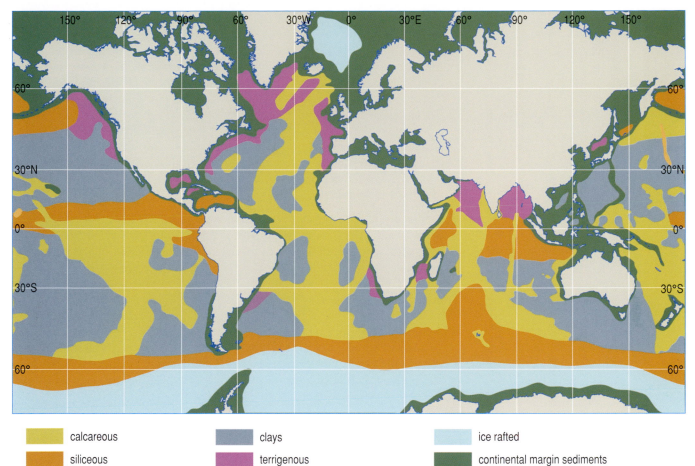

calcareous	clays	ice rafted
siliceous	terrigenous	continental margin sediments

Figure 1.12   Distribution of dominant sediment types on the floor of the present-day oceans. Note that clays are mostly also 'terrigenous'.

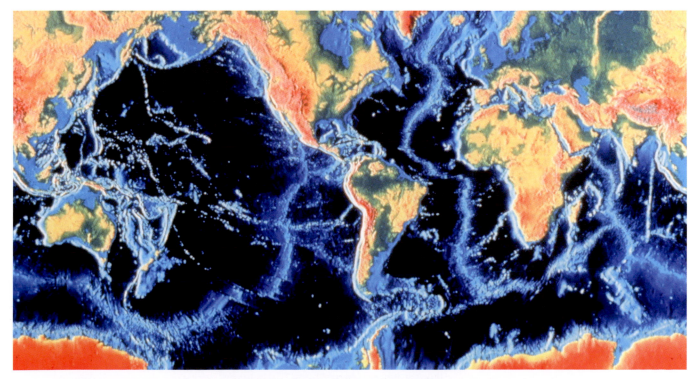

Figure 1.13   Shaded relief map of the Earth's solid surface. In oceanic areas, the deeper the blue, the deeper the water.

Clays are found over large areas of the abyssal plain, especially in the Pacific. They owe their existence as a recognizable sediment type to the *absence* of other sediment components rather than to an abundance of clay minerals as such. Thus, clays are least diluted in the deepest parts of the ocean basins, where there is a lack of biogenic material sinking from the surface, but also the chemistry of the bottom waters is such that any calcareous debris is dissolved (see Chapter 3). These regions are well away from continental margins and so the clays are less diluted by other terrigenous sediments. They tend to be brown in colour, because of oxidation of iron in the sediments, which have low accumulation rates (see Section 1.3) and experience prolonged exposure to oxygenated bottom waters on the sea-floor.

Calcareous biogenic sediments predominate along the ocean ridges which are in general the shallowest regions of the open ocean. The chemistry of near-surface waters differs from that of deep waters (see Chapter 3), such that calcareous sediments are best preserved at relatively shallow water depths. Siliceous sediments are most abundant in the Southern Ocean, where divergence of water at around 60° S causes **upwelling** which supports relatively high levels of primary production. Since diatoms are the dominant phytoplankton at low temperatures, Southern Ocean sediments are silica-rich. Siliceous sediments also occur along the Equator in the Indian and Pacific Oceans, extending north and south along the eastern Pacific margin where upwelling brings silica-rich deep water to the sea-surface.

1.3   SEDIMENT ACCUMULATION RATES

The rate at which marine sediments accumulate depends on (i) the rate of particle supply to the sea-floor, and (ii) the degree to which these particles are preserved following sedimentation. Deep-sea sediments accumulate very slowly, usually less than a few millimetres per thousand years (Table 1.2), although in areas where sediments are 'piled up' by currents, such as the Agulhas Ridge in the southern Cape Basin, sediment accumulation rates can be as high as a few tens of centimetres per thousand years. There is a small proportion of material of extraterrestrial origin in deep-sea sediments (Figure 1.14), the remains of meteorites destroyed in their passage through the Earth's atmosphere. The rate of accumulation of this meteoritic or cosmic dust in the deep sea is extremely low, and would correspond to around 0.1 to 1 mm per million years.

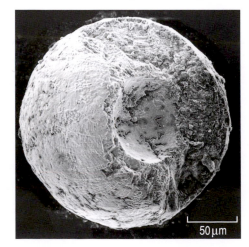

Figure 1.14   Scanning electron micrograph of a spherule of cosmic dust, typical of those found in deep-sea sediments. An iron–nickel core is surrounded by magnetite formed by oxidation during passage through the Earth's atmosphere. This spherule was found in red clay collected by HMS *Challenger* from the southern mid-Pacific. Cosmic spherules range from about 50 to 200 μm in diameter and from about 0.5 to 10 μg in weight. It has been estimated that 300×10^3 tonnes of cosmic (meteoritic) dust falls on the Earth's surface each year.

50 μm

Not surprisingly, sediments deposited on the continental margins accumulate much more rapidly, usually at rates >1 cm per 1000 years. Continental margin deposits are typically much thicker than deep-sea sediments, partly because of rapid sedimentation rates, but also because the age of the underlying crust is much older (so sediments have more time to accumulate). Young ocean crust at mid-ocean ridges has very little sediment cover.

Table 1.2 Typical sediment accumulation rates.

Area	Average sediment accumulation rate (cm 10^3 yr^{-1})
Continental margin:	
continental shelf	15–40
continental slope	20
fjord (Saanich Inlet, British Columbia)	400
Fraser River delta (British Columbia)	700 000
Enclosed/semi-enclosed seas:	
Black Sea	30
Gulf of California	100
Clyde Sea	500
Deep sea:	
coccolith ooze	0.2–3.0
clays	0.03–0.8

1.4   INTERACTIONS BETWEEN SEDIMENTS AND SEAWATER

Chemical constituents are being continually added to the ocean, largely via rivers, and we have seen that Ca^{2+}, CO_3^{2-} and SiO_2 are extracted from seawater to form skeletons of plants and animals. Other constituents are also extracted from seawater, to make the soft tissues of these plants and animals, but not all chemicals that are added to seawater are utilized for biological activity. In addition, we know that very little of the soft tissue, and not all skeletal material, survives to reach the sea-bed. Yet there is no evidence that seawater is becoming more concentrated with time, so there must be other processes going on in the oceans that act to remove constituents from solution. These processes are examined in the next Chapter.

1.5   SUMMARY OF CHAPTER 1

1   Deep-sea sediments can be classified broadly as terrigenous (land-derived) and biogenic (formed as a result of biological activity), with minor volcanogenic and cosmic contributions. Pelagic sediments include all those sediments deposited from the water column in the deep ocean basins beyond the influence of continental margin processes. Factors such as submarine topography and climatic patterns influence the type and abundance of sediments that accumulate in a particular region.

2   Pelagic biogenic sediments are composed mostly of the remains of calcareous (carbonate) and siliceous planktonic organisms, principally coccolithophores, foraminiferans, pteropods, diatoms and radiolarians.

The preservation of these depends on a number of factors such as water depth and chemistry, the shape of the skeletal remains, and the presence or absence of an organic membrane.

3   Wind-blown dusts consist mostly of clay minerals derived from weathering of terrestrial rocks. Coarse debris in deep-sea sediments is deposited by turbidity currents and ice-rafting. Deep-sea clays comprise clay minerals derived from a number of different sources. Four main types of clay minerals are recognized, each characteristic of particular weathering regimes: kaolinite, chlorite, illite and montmorillonite. Aeolian inputs are a major contribution to clays in many regions.

4   Clays predominate in the deepest parts of the ocean basins where they are not diluted by biogenic material. Calcareous sediments are largely confined to the shallowest regions of the open oceans, such as the mid-ocean ridges, while siliceous sediments predominate below regions of upwelling where there is little input from other material (i.e., terrigenous or calcareous material).

5   Accumulation rates of pelagic sediments are very slow, a few millimetres per thousand years. Sediments accumulating along continental margins do so much more rapidly, usually >1 cm per thousand years.

Now try the following questions to consolidate your understanding of this Chapter.

QUESTION 1.4   Suggest two reasons why sediments should be thicker near to continental margins than near mid-ocean ridges.

QUESTION 1.5   It has been established by careful analysis that the ratio of the concentration of Ca^{2+} to the salinity of seawater is greater in deep than in surface water. Would you expect the ratio of dissolved silica (SiO_2) to salinity to vary in the same way?

QUESTION 1.6   Explain whether you would expect to find chlorite or kaolinite in sediments: (a) of the equatorial Atlantic, (b) round Antarctica. (c) Why should illite be more common in sediments of the Northern than the Southern Hemisphere?

CHAPTER 2   BIOGEOCHEMICAL PROCESSES IN SEAWATER

'The quantity of the different elements in seawater is not proportional to the quantities of the different elements which river water pours into the sea … but inversely proportional to the facility with which elements in seawater are made insoluble by general chemical or organochemical reactions in the sea.'

Georg Forchammer (1865) *Phil. Trans. Roy. Soc.*

'When I think of the floor of the deep sea, the single, overwhelming fact that possesses my imagination is the accumulation of sediments. I see always the steady, unremitting, downward drift of materials from above, flake upon flake, layer upon layer – a drift that has continued for hundreds of millions of years, that will go on as long as there are seas and continents …. For the sediments are the materials of the most stupendous snowfall the Earth has ever seen ….'

Rachel Carson, *The Sea Around Us.*

The seawater solution primarily consists of dissolved salts, or ions, such as sodium (Na^+) and calcium (Ca^{2+}) that are largely derived from the breakdown of crustal rocks, but it also contains dissolved gases, such as oxygen (O_2), dissolved organic material, and a huge range of particles in suspension. Some of these particles are living (viruses, bacteria and phytoplankton), some are of organic origin (faecal pellets) and some inorganic (e.g. clay brought down by rivers, dust carried out to sea on the wind). The composition of the seawater solution is prone to change; for example, particles sink to the sea-floor or are advected by ocean currents, and certain ions are extracted by biological organisms and utilized for their growth.

Most of the 92 naturally occurring elements have been measured or detected in seawater, and the remainder are likely to be found as more sensitive analytical techniques become available. The elements so far determined show a vast range of concentrations (Table 2.1).

2.1   BEHAVIOUR OF DISSOLVED CONSTITUENTS

Major constituents of seawater are those ions which occur in concentrations greater than about 1 part per million (1×10^{-6}) by weight. They account for over 99.9% of the **salinity** (S) of seawater, which for our purposes is the sum of all the ions in seawater, and approximates to 35 parts per thousand by weight throughout most of the oceans. Nowadays, however, salinity is measured by electrical conductivity which is determined as a ratio relative to a standard, and so is presented correctly as numbers without units. Despite their relatively high concentrations, nitrogen and oxygen (Table 2.1) are not generally included among the major constituents because they are dissolved gases.

The remainder of the ions present in seawater are in the form of **minor and trace constituents**. Although the distinction between the two is somewhat ill-defined, for convenience here we consider minor constituents to be those with concentrations of between 1×10^{-6} and 1×10^{-9} by

Table 2.1 Average abundances of chemical elements in seawater.

Element		Concentration (mg l^{-1})	Some probable dissolved species	Total amount in the oceans (tonnes)
chlorine	Cl	1.95×10^4	Cl^-	2.57×10^{16}
sodium	Na	1.077×10^4	Na^+	1.42×10^{16}
magnesium	Mg	1.290×10^3	Mg^{2+}, $MgSO_4$, $MgCO_3$	1.71×10^{15}
sulphur	S	9.05×10^2	SO_4^{2-}, $NaSO_4^-$	1.20×10^{15}
calcium	Ca	4.12×10^2	Ca^{2+}	5.45×10^{14}
potassium	K	3.80×10^2	K^+	5.02×10^{14}
bromine	Br	67	Br^-	8.86×10^{13}
carbon	C	28	HCO_3^-, CO_3^{2-}, CO_2	3.70×10^{13}
nitrogen	N	11.5	N_2 gas, NO_3^-, NH_4^+	1.50×10^{13}
strontium	Sr	8	Sr^{2+}	1.06×10^{13}
oxygen	O	6	O_2 gas	7.09×10^{12}
boron	B	4.4	$B(OH)_3$, $B(OH)_4^-$, $H_2BO_3^-$	5.82×10^{12}
silicon	Si	2	$Si(OH)_4$	2.64×10^{12}
fluorine	F	1.3	F^-, MgF^+	1.72×10^{12}
argon	Ar	0.43	Ar gas	5.68×10^{11}
lithium	Li	0.18	Li^+	2.38×10^{11}
rubidium	Rb	0.12	Rb^+	1.59×10^{11}
phosphorus	P	6×10^{-2}	HPO_4^{2-}, PO_4^{3-}, $H_2PO_4^-$	7.93×10^{10}
iodine	I	6×10^{-2}	IO_3^-, I^-	7.93×10^{10}
barium	Ba	2×10^{-2}	Ba^{2+}	2.64×10^{10}
molybdenum	Mo	1×10^{-2}	MoO_4^{2-}	1.32×10^{10}
uranium	U	3.2×10^{-3}	$UO_2(CO_3)_2^{4-}$	4.23×10^9
vanadium	V	2×10^{-3}	$H_2VO_4^-$, HVO_4^{2-}	2.64×10^9
arsenic	As	2×10^{-3}	$HAsO_4^{2-}$, $H_2AsO_4^-$	2.64×10^9
nickel	Ni	4.8×10^{-4}	Ni^{2+}, $NiCO_3$, $NiCl^+$	6.35×10^8
zinc	Zn	4×10^{-4}	$ZnOH^+$, Zn^{2+}, $ZnCO_3$	5.29×10^8
aluminium	Al	4×10^{-4}	$Al(OH)_4^-$	5.29×10^8
caesium	Cs	4×10^{-4}	Cs^+	5.29×10^8
chromium	Cr	3×10^{-4}	$Cr(OH)_3$, CrO_4^{2-}	3.97×10^8
antimony	Sb	2×10^{-4}	$Sb(OH)_6^-$	2.64×10^8
krypton	Kr	2×10^{-4}	Kr gas	2.64×10^8
selenium	Se	2×10^{-4}	SeO_3^{2-}, SeO_4^{2-}	2.64×10^8
neon	Ne	1.2×10^{-4}	Ne gas	1.59×10^8
cadmium	Cd	1×10^{-4}	$CdCl_2$	1.32×10^8
copper	Cu	1×10^{-4}	$CuCO_3$, $CuOH^+$, Cu^{2+}	1.32×10^8
tungsten	W	1×10^{-4}	WO_4^{2-}	1.32×10^8
xenon	Xe	5×10^{-5}	Xe gas	6.61×10^7
iron	Fe	2.2×10^{-5}	$Fe(OH)_2^+$, $Fe(OH)_4^-$	2.90×10^7
manganese	Mn	2×10^{-5}	Mn^{2+}, $MnCl^+$	2.64×10^7
yttrium	Y	2×10^{-5}	$Y(OH)_3$, YCO_3^+, Y^{3+}	2.64×10^7
zirconium	Zr	1×10^{-5}	$Zr(OH)_5^-$	1.32×10^7
thallium	Tl	1×10^{-5}	Tl^+	1.32×10^7
thorium	Th	1×10^{-5}	$Th(OH)_4$	1.32×10^7
hafnium	Hf	7×10^{-6}	$Hf(OH)_5^-$	9.25×10^6
helium	He	6.8×10^{-6}	He gas	8.99×10^6
titanium	Ti	6.5×10^{-6}	$Ti(OH)_4$	8.59×10^6
germanium	Ge	5×10^{-6}	$Ge(OH)_4$, $H_3GeO_4^-$	6.61×10^6
rhenium	Re	4×10^{-6}	ReO_4^-	5.29×10^6
cobalt	Co	3×10^{-6}	Co^{2+}	3.97×10^6
lanthanum	La	3×10^{-6}	$La(OH)_3$	3.97×10^6
neodymium	Nd	3×10^{-6}	$Nd(OH)_3$	3.97×10^6
lead	Pb	2×10^{-6}	$PbCO_3$, $Pb(CO_3)_2^{2-}$	2.64×10^6
silver	Ag	2×10^{-6}	$AgCl_2^-$	2.64×10^6

Element		Concentration (mg l^{-1})	Some probable dissolved species	Total amount in the oceans (tonnes)
gallium	Ga	2×10^{-6}	$Ga(OH)_4^-$	2.64×10^6
cerium	Ce	1×10^{-6}	$Ce(OH)_3$, $CeCO_3^+$, Ce^{3+}	1.32×10^6
dysprosium	Dy	9×10^{-7}	$Dy(OH)_3$, $DyCO_3^+$, Dy^{3+}	1.19×10^6
erbium	Er	8×10^{-7}	$Er(OH)_3$, $ErCO_3^+$, Er^{3+}	1.06×10^6
ytterbium	Yb	8×10^{-7}	$Yb(OH)_3$, $YbCO_3^+$, Yb^{3+}	1.06×10^6
gadolinium	Gd	7×10^{-7}	$Gd(OH)_3$, $GdCO_3^+$, Gd^{3+}	9.25×10^5
tin	Sn	6×10^{-7}	$SnO(OH)_3^-$	7.93×10^5
praseodymium	Pr	6×10^{-7}	$Pr(OH)_3$, $PrCO_3^+$, Pr^{3+}	7.93×10^5
scandium	Sc	6×10^{-7}	$Sc(OH)_3$	7.93×10^5
samarium	Sm	6×10^{-7}	$Sm(OH)_3$	7.93×10^5
holmium	Ho	3×10^{-7}	$Ho(OH)_3$, $HoCO_3^+$, Ho^{3+}	3.97×10^5
mercury	Hg	3×10^{-7}	$HgCl_4^{2-}$, $HgCl_2$	3.97×10^5
niobium	Nb	3×10^{-7}	$Nb(OH)_6^-$	3.97×10^5
lutetium	Lu	2×10^{-7}	$Lu(OH)$	2.64×10^5
thulium	Tm	2×10^{-7}	$Tm(OH)_3$	2.64×10^5
beryllium	Be	2×10^{-7}	$BeOH^+$	2.64×10^5
europium	Eu	2×10^{-7}	$Eu(OH)_3$, $EuCO_3^+$, Eu^{3+}	2.64×10^5
indium	In	1×10^{-7}	$In(OH)_2^+$, $In(OH)_3$	1.32×10^5
terbium	Tb	1×10^{-7}	$Tb(OH)_3$, $TbCO_3^+$, Tb^{3+}	1.32×10^5
rhodium	Rh	8×10^{-8}	$Rh(OH)_3$?	1.06×10^5
tellurium	Te	8×10^{-8}	$Te(OH)_6$	1.06×10^5
palladium	Pd	5×10^{-8}	Pd^{2+}, $PdCl^+$	6.61×10^4
platinum	Pt	5×10^{-8}	$PtCl_4^{2-}$	6.61×10^4
tantalum	Ta	4×10^{-8}	$Ta(OH)_5$	5.28×10^4
bismuth	Bi	2×10^{-8}	BiO^+, $Bi(OH)_2^+$	2.64×10^4
gold	Au	2×10^{-8}	$AuCl_2^-$	2.64×10^4
osmium	Os	9×10^{-9}	$H_3OsO_6^{2-}$?	1.18×10^4
ruthenium	Ru	2×10^{-9}	not known	2.64×10^3
iridium	Ir	1.3×10^{-10}	$Ir(OH)_3$	172
radium	Ra	7×10^{-11}	Ra^{2+}	92.5
protactinium	Pa	5×10^{-11}	not known	66.1
radon	Rn	6×10^{-16}	Rn gas	7.93×10^{-4}
polonium	Po	5×10^{-16}	Po_3^{2-}, $Po(OH)_2$?	6.61×10^{-4}

IMPORTANT:

1 Table 2.1 does not represent the last word on seawater composition. Even for the most abundant constituents, compilations from different sources differ in detail. For the rarer elements, many of the entries in Table 2.1 will be subject to revision, as analytical methods improve and more data become available.

2 Concentrations in Table 2.1 are by weight of the element. While this is convenient for some purposes, for many others it is more useful to express concentrations in molar terms: one mole of any element (or compound) has a mass in grams equal to the atomic (or ionic, or molecular) mass of the element (or compound). Thus, a mole of calcium contains 40 g; a mole of magnesium contains 24 g, a mole of carbonate ions (CO_3^{2-}) contains $12 + (16 \times 3) = 60$ g; and so on.

3 Concentrations of gases are given in mg l^{-1} in Table 2.1 and these are numerically not very different from the volumetric concentrations (ml l^{-1}) for oxygen, nitrogen, argon (and some other gases). That is because of their low densities: 1.43, 1.25 and 1.78 kg m^{-3} respectively. (So, to a first approximation, 1 m^3 weighs 1000 g, 1 litre weighs 1 gram, and 1 ml weighs 1 mg.)

4 *Nitrogen and nitrate:* Dissolved nitrogen gas as N_2 is biologically almost inert and generally participates very little in marine biological processes. Only small amounts of it are fixed (i.e. incorporated into living tissue) by micro-organisms (the so-called blue–green algae or **Cyanobacteria**). The total concentration of nitrogen in average ocean water is given as 11.5 mg l^{-1} in Table 2.1. Only about 0.5 mg l^{-1} of that total is in a form other than N_2 gas dissolved from the atmosphere, and this small proportion (i.e. 0.5 mg l^{-1}) participates in marine biological cycles. That is because it consists of fixed (i.e. chemically combined) nitrogen, mainly as nitrate or ammonium ions.

5 There is a wide variety of particulate matter in the oceans, and the distinction between what constitutes material truly in solution and what is particulate matter can present problems in the analysis of seawater. Filtration through a membrane having pores of diameter 0.2 μm is a widely used procedure for separating dissolved from particulate fractions. This is satisfactory for most constituents, but it is important to remember that the 'dissolved' fraction also includes **colloidal particles**; these are tiny particles that remain in suspension indefinitely (e.g. $Fe(OH)_3$).

weight, and trace constituents to be those with concentrations of less than 1×10^{-9} by weight (1 part per billion).

Apart from the obvious distinction between major constituents and the rest in terms of mass, there is another reason for distinguishing between them. Salinity values can vary from less than 33 to more than 37 throughout the oceans, depending on the extent to which freshwater is added by precipitation, run-off, and melting ice and snow, or removed by evaporation. But the ratio of the concentrations of most of the individual major dissolved constituents to total salinity remains practically constant. This **constancy of composition of seawater** is maintained because most of the major constituents exhibit **conservative** behaviour, that is, their concentrations in seawater are not significantly changed by the biological and chemical reactions that take place within the main body of the oceans. Their concentrations can be changed only by mixing between different water masses of contrasted salinity. By contrast, nearly all of the minor and trace dissolved constituents exhibit **non-conservative** behaviour: their concentrations are significantly changed by biological and chemical reactions in seawater.

From your reading of Chapter 1, which of the three major dissolved constituents must be exceptions to the generalization above, in that they are non-conservative?

They are carbon, calcium and silicon. In fact, Ca^{2+} is so abundant in seawater that the ratio of its concentration to total salinity (the $Ca : S$ ratio) is only very slightly greater in deep than in surface water, and Ca^{2+} departs only to a small degree from strictly conservative behaviour. The $C : S$ ratio is more variable; there is more carbon in deep than in surface waters, and there is also some inter-ocean variability. The $Si : S$ ratio varies greatly, both with depth and from place to place. Because of this strongly non-conservative behaviour, silicon (as dissolved silica, SiO_2) is not generally classified among major constituents, even though the average concentration of Si is above the $1 \, mg \, l^{-1}$ limit in Table 2.1.

A small number of minor and trace dissolved constituents also behave conservatively, but the great majority are non-conservative. Their concentrations can change considerably from place to place, but as those concentrations are so low (Table 2.1), such variations have no detectable effect either on total salinity or on the overall constancy of composition of seawater.

QUESTION 2.1 (a) How does Figure 2.1 suggest at first sight that nitrate and barium participate in biological cycles and sodium does not?

(b) Which of the profiles demonstrate(s) (i) conservative, (ii) non-conservative behaviour?

(c) Sodium is an example of an element that is essential to life, and is clearly not biologically 'inert'. How can this be reconciled with your answers to (a) and (b)?

At this point, it is necessary to emphasize that the distinction between conservative and non-conservative behaviour depends on the extent to which an ion participates in biological or chemical reactions, *in relation to its overall concentration*. For example, small amounts of both Mg^{2+} and Sr^{2+} can follow Ca^{2+} into the carbonate skeletons of some marine organisms, and a minority of zooplankton (including some radiolarians)

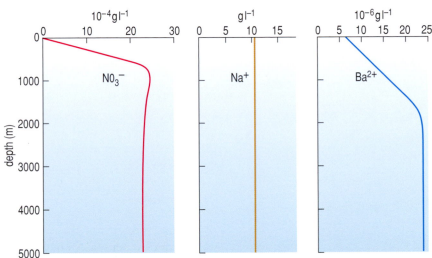

Figure 2.1 Profiles for three seawater constituents. Concentrations are normalized to a salinity of 35. For discussion, see Question 2.1 and related text.

can even secrete skeletons of celestite, strontium sulphate ($SrSO_4$). However, the amounts of Mg^{2+} and Sr^{2+} involved in such transactions are so small in relation to their total quantity in the oceans that the effects are barely detected analytically. Mg^{2+} is generally grouped among the conservative constituents, Sr^{2+} among the non-conservative ones (but only just, as in the case of Ca^{2+}).

2.1.1 THE STEADY-STATE OCEAN

The consensus among marine scientists is that the oceans are chemically in a **steady state**, at least as far as the major dissolved constituents of seawater are concerned, and probably in the case of many of the minor and trace constituents as well. This means that the chemical budgets balance, such that the rate of supply of dissolved constituents equals the rate of removal. The available evidence suggests that the steady-state condition may be a characteristic feature of the oceans over periods of millions of years.

The concept of the steady-state ocean allows us to define a mean oceanic **residence time** (τ), which is given by:

$$\tau = \frac{\text{total mass of substance dissolved in the oceans}}{\text{rate of supply (or removal) of the substance}}$$

Residence times of most elements are long compared with the average oceanic mixing time of about 500 years, so the majority of dissolved constituents should be uniformly distributed throughout the oceans. This is true for conservative constituents which have the longest residence times (more than 10^5 years). Non-conservative constituents are not uniformly distributed — concentrations vary with depth (Figure 2.1) and from place to place, even though many have residence times that are much longer than the mean oceanic mixing time.

QUESTION 2.2 Explain why a constituent with an oceanic residence time of 100 years cannot be uniformly mixed throughout the oceans.

Even the shortest residence times, however, are long compared with the lifespan of most marine organisms, so there is ample opportunity for dissolved constituents to participate in biological cycles several times over.

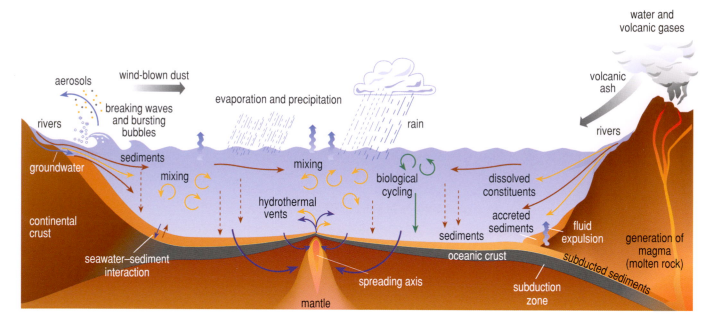

Figure 2.2 Diagrammatic cross-section illustrating the role of the ocean in the global cycling of elements through mantle, crust, rivers, atmosphere and oceans.

All of this leads us to the oceanic biogeochemical cycle; this is illustrated schematically in Figure 2.2 and summarized below:

1   Terrigenous sediments are supplied to the oceans by rivers, by winds, and from volcanic eruptions (meteorite infall is sufficiently small to be ignored here). Dissolved constituents are supplied mainly by rivers, but inputs from **hydrothermal** fluids (formed during high-temperature reactions between seawater and newly formed oceanic crust, especially at ocean ridge **spreading axes**) are important for some constituents, like manganese. Though difficult to quantify, discharge of terrestrial **groundwaters** to the coastal zone may also be important for some constituents, including the nutrient elements and calcium. There are also inputs from volcanic gases, and there is continuous exchange of gases between atmosphere and ocean across the air–sea interface.

2   Of the approximately 4×10^9 tonnes of dissolved material entering the oceans from rivers each year, about 10% consists of **cyclic salts**. Oceanic **aerosols**, formed by breaking waves and bursting bubbles, contain salts which are dispersed throughout the atmosphere and recycled back to the oceans by rain and rivers. Much of the chloride, nearly half of the sodium and significant amounts of other major dissolved constituents entering the oceans from rivers, are contributed as cyclic salts.

3   The majority of the dissolved constituents participate in repeated cycles of biological activity that remove elements from solution and return them to solution many times, before they end up in sediments. Some constituents (e.g. sulphate and magnesium) are also removed directly into oceanic crust during hydrothermal activity.

4   At and below the sea-bed itself, there are chemical exchanges between seawater and sediments (see Chapter 5) as well as between seawater and igneous rocks of the underlying oceanic crust. The latter processes include both low-temperature **sea-floor weathering** of basaltic rocks and the high-temperature hydrothermal reactions mentioned in item 1 above. Reactions that remove dissolved constituents from seawater are sometimes called reactions of *reverse weathering*, because some of them, especially those

forming calcareous and siliceous sediments, are essentially the reverse of the reactions that supply the original constituents during continental weathering.

5 The oceans are in compositional steady state because the rates of supply of dissolved constituents from the various *sources* (items 1 and 2 above) are balanced by the rates of removal of these constituents into the *sinks* (items 3 and 4 above).

6 Ultimately, oceanic crust is carried back down into the Earth's mantle at **subduction zones**. Some of the overlying sediments (plus seawater trapped in them) are scraped off and accreted to continental margins, and some are carried into the mantle. Both processes ensure that much of the material eventually finds its way back into the cycle of weathering and erosion at the Earth's surface and so eventually back into the oceans.

We should not leave Figure 2.2 without mentioning **excess volatiles**. Mass balance calculations show that for several dissolved constituents of seawater, concentrations in crustal rocks are much too low for continental weathering to account for their abundance in the oceans. A good example is chloride, the most abundant element in the seawater solution (Table 2.1): its concentration in average crustal rock is 0.01%. Sulphur, boron and bromine are other major elements which are classed as excess volatiles and there are several minor and trace constituents also. All are found in volcanic gases, and the bulk of these excess volatiles may have been supplied to the oceans by volcanoes early in the Earth's history. They continue to be expelled in volcanic gases at the present time: it has been estimated that volcanic eruptions expel something like one million tonnes of chloride (as HCl) into the atmosphere each year (a minute amount compared to the total mass of chloride in the oceans, Table 2.1), but this is being augmented by increased inputs from anthropogenic sources.

In the past few decades, it has become apparent that a major factor in the behaviour of many dissolved constituents, and hence in the regulation of seawater composition, is the oceanic particle cycle. Particles produced by biological processes in surface water sink into the deep ocean and participate in a variety of biogeochemical interactions on the way. Thus, as noted in Section 1.4, the relative proportions of individual elements making up the sediments on the deep sea-floor differ considerably from those in the original particles produced in surface waters.

2.2 THE BIOLOGICAL PARTICLE CYCLE

Most of the particulate matter falling from the surface layers of the ocean is produced initially by *primary producers*, the photosynthesizing **phytoplankton** (the 'grass of the ocean') in the sunlit **photic zone**, which rarely extends more than 150 m below the surface and generally much less. Some of the phytoplankton die naturally, but most are grazed by **zooplankton**, small planktonic animals that package most of their waste products into faecal pellets. The faecal pellets are in turn consumed and decomposed by other organisms, including bacteria. Thus, most biological material is recycled in the surface layers, but a small proportion survives in the particles that sink out of the photic zone

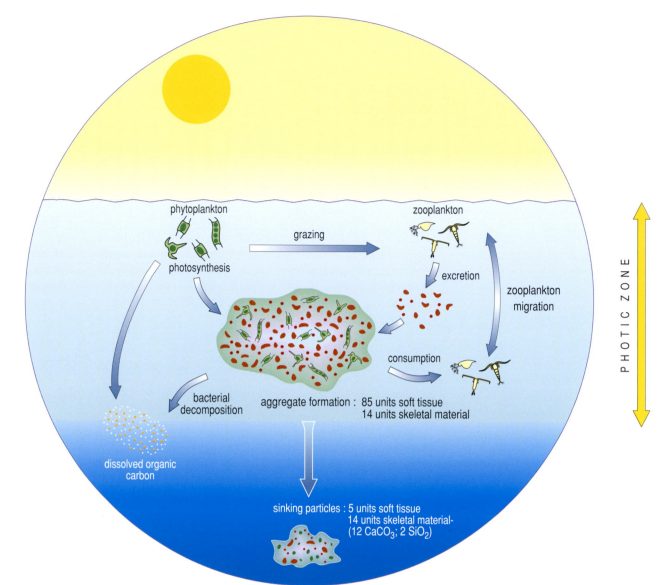

phytoplankton

zooplankton

grazing

photosynthesis

excretion

zooplankton migration

consumption

bacterial decomposition

aggregate formation : 85 units soft tissue
14 units skeletal material

dissolved organic carbon

PHOTIC ZONE

sinking particles : 5 units soft tissue
14 units skeletal material
(12 CaCO₃; 2 SiO₂)

Figure 2.3   Simplified diagram of biological particle formation in the oceans, showing approximate average make-up of the particulate material. The high degree of recycling of organic matter in the photic zone ensures that on average it comprises only about one-quarter of the material sinking into deep water. (Note that of the 14 units of skeletal material, on average 12 units are CaCO₃, 2 units are SiO₂.)

towards the sea-bed (Figure 2.3). Figure 2.3 distinguishes between two different types of biological material. First, there are the soft tissues, which are rapidly broken down on the death of the organism. Secondly, there is the skeletal component which comprises the hard parts of organisms, such as shells and skeletons. By convention, the soft tissues are referred to as the 'organic' component of the biological particulate material; this is because the shells and skeletons, although they are produced by organisms (and are thus 'organic'), have a mineral structure and do not decompose rapidly on the death of the organism.

As biological particles fall through the water column, the organic component will not only provide food for successive populations of filter-feeders and other animals, so that it is repackaged several times *en route* to the sediment, but will also carry its own communities of microbial decomposer organisms.

The principal chemical constituents that make up the soft tissues of *all* organisms are: oxygen, hydrogen, carbon, nitrogen and phosphorus. Of these, nitrogen and phosphorus are the two most important. This is because they are not always available in forms that are biologically utilizable, so (in addition to light) they can potentially limit phytoplankton productivity. In contrast, oxygen, hydrogen and carbon are abundant everywhere. Those organisms that form hard parts utilize additional elements; foraminiferans require carbon and calcium, while radiolarians and diatoms require silica.

About three-quarters of organic matter in the biological particles that leave the photic zone is decomposed and recycled in the upper 500–1000 m of the water column, i.e. above the main (or permanent) **thermocline**; and, on average, only about 1% of the sinking organic matter actually reaches the sediments. In other words, of the five units that contribute to the sinking particles in Figure 2.3, three or four units are recycled above the thermocline, and on average only about 0.05 unit survives to reach the sediments; and so organic-rich sediments are rare in the deep oceans. By contrast, the proportion of skeletal material increases as particles sink deeper because the chemical processes whereby it is dissolved in general operate more slowly than biological processes. *Thus, the average composition of the biological particles changes with depth such that the proportion of skeletal and inorganic components increases greatly.* Before looking at these processes in more detail, we first examine the nature of the particulate matter itself.

2.2.1 PARTICULATE ORGANIC MATTER

As already noted, almost all particulate matter in seawater is generated *in situ* by primary producers, such as phytoplankton. Phytoplankton range in size from less than 1 μm to a few tens of μm — so small that, theoretically, they should take hundreds, or even thousands, of years to settle through the water column after death. Why, then, are the calcite shells of tiny coccoliths preserved on the deep ocean floor beneath the area they were produced, not carried far from their source by currents, or dissolved during a trip to the bottom lasting several centuries? The reason is the repackaging mentioned earlier; the fine, light particles do not settle individually but are repackaged into larger particles, up to a few hundreds of μm, that settle to the deep sea at a much greater speed. The products of this **biological aggregation** include faecal pellets from zooplankton. Filter-feeding zooplankton are concerned only with the size of their food and graze phytoplankton almost indiscriminately, so indigestible coccoliths and diatom frustules are concentrated in their faecal pellets (Figure 2.4). These act as ballast, and faecal pellets sink as rapidly as 100 to 200 m per day.

Figure 2.4 A transmission electron microscope image (enlarged × 3000) of a section of a faecal pellet collected by a sediment trap in the North Atlantic. It includes undigested phytoplankton cells. The blue (artificially coloured) areas represent remains of coccoliths that were eaten by zooplankton and passed unaffected through the gut. Coccoliths are composed of nearly pure calcite, the heaviest mineral produced by marine organisms, which makes a faecal pellet heavy enough to settle rapidly through the water column.

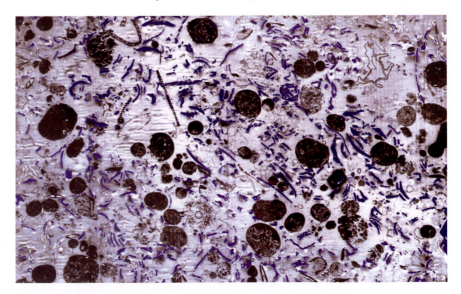

Figure 2.5   A marine snow particle of diameter 4 mm. This specimen, from the Atlantic Ocean, comprises dead and decaying phytoplankton, zooplankton faecal matter and exoskeletons.

Examination of sediment trap material has allowed us to learn much more about repackaging (Box 2.1).

In addition, there are easily visible (macroscopic) aggregates known as **marine snow** (Figure 2.5), consisting of detritus, living organisms (including bacteria) and some inorganic matter (mostly clay mineral particles). Zooplankton produce web-like material that is sticky and fibrous onto which small particles such as faecal pellets, foraminiferan tests and airborne dust can easily adhere, forming aggregates. Eventually, the aggregate becomes heavy enough to sink. As it descends, more suspended particles are added, making the aggregate even heavier and faster moving. The biological 'glues' that bind the aggregates are not especially strong, so marine snow is continually disintegrating and re-forming, shedding some of its components and gaining new ones as it sinks towards the sea-bed.

Most marine snow, however, never makes it to the sea-bed. Most is disaggregated and/or eaten and repackaged into faecal pellets in the upper 500–1000 m of the water column, that is, above the permanent thermocline. Below about 1000 m, faecal pellets provide the main component of the sinking particle flux.

In some circumstances, however, the rate of production of marine snow can exceed the rate of disaggregation and/or consumption in the upper water column.

What might those circumstances be?

Strong seasonal blooms of phytoplankton may lead to formation of marine snow in such large quantities that significant amounts of organic matter can reach the sea-bed more or less intact, even in water depths of 4000 m or more. For example, in the Arabian Sea, there is a strong correlation between primary production in surface waters and the organic carbon content of particles recovered from the water column at about 3.5 km depth (Figure 2.6). This means that particles must have sinking velocities of a few hundred metres per day, reaching the sea-bed in a matter of a few days. It has been

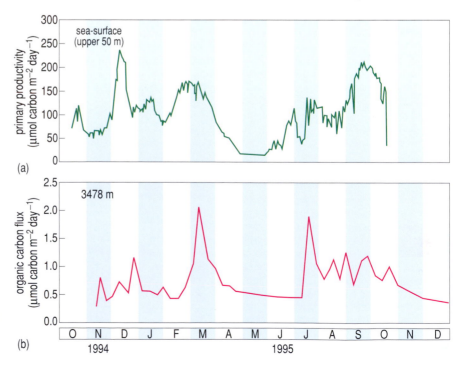

Figure 2.6   Time series plots of (a) surface primary productivity and (b) organic carbon flux measured by a sediment trap at 3478 m below the sea-surface, for a site in the Arabian Sea.

BOX 2.1   CATCHING PARTICLES IN THE ABYSS

Much of our information about the nature and distribution of particulate organic matter in the water column comes from material collected in **sediment traps** — the oceanographer's equivalent of rain gauges, which they rather resemble in appearance (Figure 2.7). Sediment traps are usually deployed on moorings, suspended above the sea-floor and reaching to within about 500 m of the sea-surface.

Sediment traps are usually deployed for about a year. The sample bottles located at the bottom of the funnel can be rotated at intervals in order to collect a series of samples (Figure 2.8). Scientists can programme the period of collection of an individual bottle before deployment; it is common to sample more frequently during an expected bloom season, and to sample for longer in winter when production is low. At the end of the collection period, the bottle is closed and the next bottle opens.

Sampling using sediment traps is not straightforward. The trap must be kept stable, maintaining its upright position at all times during deployment, and it must also be retrieved intact. There are also questions about whether a trap collects all of the particles settling through the water column, or whether it can be assumed that the particles came from directly above: horizontal flow past a trap can set up vortices that can either enhance or reduce collection of sediment, depending upon the settling velocity of the particles, the speed of the current, and the tilt and geometry of the trap. New types of sediment traps that drift freely with the currents, so that the flow past the trap is effectively zero, are now being tested.

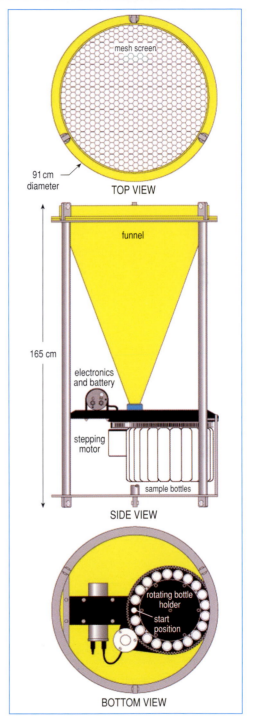

Figure 2.8   Design of a sediment trap and sample bottles. The mesh screen is designed to prevent large fish etc. from entering the funnel. Sample bottles rotate on a schedule programmed before deployment, each collecting about 5 days' to a month's worth of particulate material before being sealed for analysis on recovery of the trap. The sample bottles may contain a preservative, such as formalin, to keep organic matter samples in the best possible condition until they can be retrieved (and prevent organisms such as zooplankton from feeding on them!).

Figure 2.7   Launching a sediment trap mooring in the Indian Ocean.

suggested that early in the growth season algal flocs (fluffy aggregates) are the most important way in which organic matter is exported to the sea-bed from surface waters, while faecal transport becomes important later as zooplankton increase in abundance. In general terms, marine snow is mostly found where there is high biological productivity, which is usually seasonal. In regions of lower productivity, faecal pellets make up most of the sinking particle flux.

The vertical export of organic carbon and associated elements from the surface ocean into deeper waters is often referred to as the '**biological pump**'. A proper understanding of the operation of this pump has become increasingly important because export of biogenic carbon from the upper ocean leads to draw-down of atmospheric carbon dioxide; the degree to which the oceans can mitigate the build-up of greenhouse gases is therefore dependent on the efficiency of this pump.

Dissolved organic matter

There is as an order of magnitude more organic carbon in dissolved form in the oceans than there is organic carbon in particles. The bulk of this consists of compounds with large relative molecular masses that are unavailable for biological consumption and extremely resistant to degradation (i.e. they are **refractory**).

Compounds of low relative molecular mass represent some 0–6% of **dissolved organic carbon** (**DOC**) in the open ocean. These tend to be metabolic by-products, especially of the phytoplankton, and include amino acids, vitamins and sugars. They provide the main substrate for bacterial growth so are rapidly recycled in the upper ocean on time-scales of minutes to days. It is now recognized, however, that a tiny fraction of these low molecular mass compounds may escape microbial degradation in surface waters long enough to be mixed downwards, contributing to carbon export to the deep ocean (the biological pump). An example of this occurs in the northern North Atlantic, a region of deep-water formation. The deep waters form from DOC-rich ($\sim$60 μmol kg^{-1}) surface waters that originate from the subtropics. DOC concentrations decrease along the path of deep-water flow from 48 μmol kg^{-1} in the Greenland Sea to 41 μmol kg^{-1} in the South Atlantic (Figure 2.9); this is partly due to mixing with DOC-poor water, but it is also due to continuing bacterial respiration of material that escaped degradation in surface waters.

In situations such as this, where concentrations of DOC are higher at the surface than in deeper waters, and vertical mixing is rapid, DOC can represent a significant fraction of carbon export out of the photic zone (around 20–40%). For the ocean as a whole, however, the average is more likely to be around 10%.

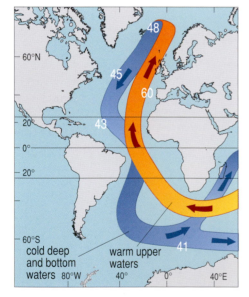

Figure 2.9   A map of DOC concentrations in the Atlantic Ocean in relation to formation of deep water in the North Atlantic. Numbers are concentrations in units of μmol kg^{-1}. The orange and blue 'ribbons' are a symbolic representation of the overall effect of the thermohaline conveyor; warm surface waters (orange arrows, orange ribbon) are transported northwards towards Greenland, where they are cooled, become more dense and sink. The blue ribbon and blue arrows show the subsequent transport of this cold water at depth throughout the Atlantic.

Not all of the organic matter in seawater (whether particulate or dissolved) is produced there. Some comes from terrestrial sources, transported to the sea by rivers and winds in particulate forms, and by rivers in dissolved forms. The latter include humic substances produced in terrestrial waters by reactions occurring between molecular fragments derived from land plants. Organic matter of terrestrial origin may locally be more abundant than organic matter of marine origin in coastal and shelf waters; but most of the particulate component is consumed or deposited there, and in global terms little survives to be exported to the deep sea. Particulate organic matter of terrestrial origin makes a small contribution to the annual aeolian supply to the oceans of *c*. 10^8 tonnes (Section 1.1.2).

2.2.2 PARTICULATE ORGANIC MATTER AND THE NUTRIENT CYCLE

The production of particulate organic matter in the photic zone, its breakdown and transport to the deep ocean, are not only important for the carbon cycle, but also for other elements that are required in the processes of living organisms. Quantitatively, the three most important **nutrient** elements are fixed nitrogen (chiefly as nitrate, NO_3^-), phosphorus as phosphate (PO_4^{3-}) and, for those species that require it for construction of their skeleton, dissolved silica (SiO_2 for brevity, but mainly as $Si(OH)_4$, Table 2.1).

These nutrients are heavily utilized in the photic zone, where their availability can limit primary production, and they can be almost totally depleted in surface waters. Consumption and decomposition of organic matter sinking from surface waters return the nutrients to solution. As a result, typical profiles look like Figure 2.10.

QUESTION 2.3 (a) Can you suggest why well-stratified surface waters are more likely to be rapidly depleted in nutrients than a well-mixed water column?

(b) Why do profiles for nitrate and phosphate reach maxima at about 1 km depth, while the maximum for silica is reached somewhat deeper?

Minor and trace constituents of seawater (hereafter both referred to as trace elements for convenience) may also be utilized by marine organisms in various ways. For example, copper and vanadium are incorporated in the blood pigments of molluscs and ascidians (sea-squirts); some sponges accumulate titanium; cobalt is used to synthesize vitamin B12; zinc is a constituent of many enzymes; and iron is an important constituent of the blood in many animals (in haemoglobin) as well as being an essential element for plants.

Trace elements that are incorporated either into soft tissues or into skeletal material will become more or less depleted in surface waters and enriched

Figure 2.10   Typical profiles in subtropical and tropical waters for concentrations of (a) dissolved phosphate, (b) dissolved nitrate, (c) dissolved silica. All concentrations are in molar terms.

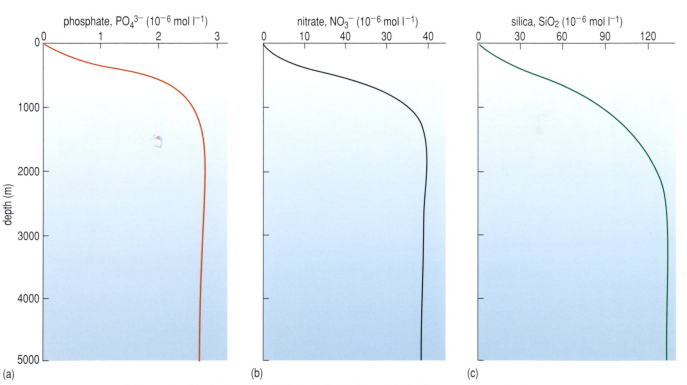

in the deep ocean. They are classified as **recycled elements** and most of them have residence times of less than 10^6 years.

QUESTION 2.4 The distribution of cadmium has been found to be well correlated with that of phosphate in the oceans. What might its concentration profile look like? You are *not* expected to put scales on your sketch.

Dissolved constituents in the recycled category can be classified as **biolimiting** or **biointermediate**, according to whether or not their availability in surface waters can limit primary production. The nutrients represented in Figure 2.10 are thus biolimiting, while carbon, calcium and barium (Figure 2.1) are bio-intermediate. Convenient though this classification is, however, it should be treated with caution, because it carries the implication that production will cease only when one or more constituents conventionally defined as biolimiting are exhausted. This raises the question of how many potentially biolimiting constituents there might be in seawater. For example, research in both Antarctic and equatorial waters has shown that primary production can cease when surface waters become depleted in iron, even though supplies of nitrate and phosphate are still available (Box 2.2).

BOX 2.2   THE IRON HYPOTHESIS

Over 20% of the world's open ocean surface waters are replete with light and the major plant nutrients (nitrate, phosphate and silica), yet **standing stocks** of phytoplankton (measured as the amount of chlorophyll per unit volume of seawater) remain low. One idea is that phytoplankton growth in these so-called high-nitrate low-chlorophyll (**HNLC**) regions (Figure 2.11) is limited by iron. Most of the iron in normal seawater exists in its oxidized Fe(III) form, which is very insoluble and therefore unavailable for biological consumption.

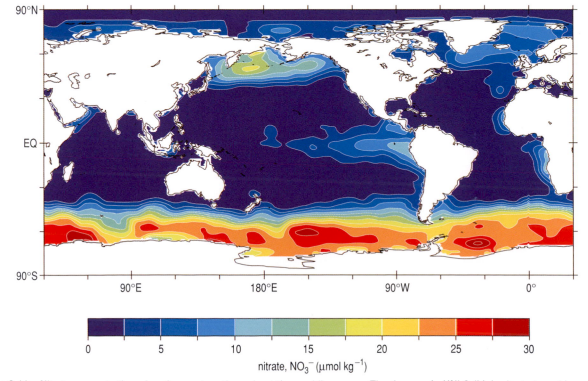

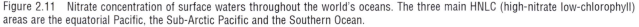

Figure 2.11   Nitrate concentration of surface waters throughout the world's oceans. The three main HNLC (high-nitrate low-chlorophyll) areas are the equatorial Pacific, the Sub-Arctic Pacific and the Southern Ocean.

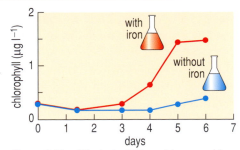

Figure 2.12   Effect of addition of iron on chlorophyll levels in surface seawater taken from a HNLC area.

Direct evidence for iron limitation was first demonstrated in experiments performed by John Martin and colleagues at Moss Landing Marine Laboratories in California. Bottles filled with surface waters from HNLC regions were stored under conditions of light and temperature equivalent to those from which the water was collected. Iron was added to half of the bottles, and the other half were left alone as controls. The general results were always the same: after one week, the amount of chlorophyll in the iron-enriched bottles was higher than in the control bottles (Figure 2.12), and the concentration of nitrate was always lower in the water in the iron-enriched bottles than in the control bottles. Not all scientists readily accepted these results, however. They pointed out that the water samples did not contain larger zooplankton, which are not sufficiently abundant to be captured by the water samplers, thereby altering the functioning of the food web in the bottles. The next step then, was to enrich a patch of the ocean itself with iron to see how the phytoplankton would respond when the food web was intact.

The first open-ocean iron enrichment experiment was conducted in the equatorial Pacific Ocean in 1993. 450 kg of dissolved iron (Fe(II)) were added to an area approximately 60 km^2, increasing iron concentrations in surface waters by a factor of 40. Phytoplankton production (particularly of the larger algae, e.g. diatoms) increased four-fold (Figure 2.13a). Further experiments in this area, and also in the Southern Ocean (Figure 2.13b), have now shown unequivocally that phytoplankton growth in these HNLC areas is limited by iron availability.

Iron fertilization of the ocean is a hot topic because stimulation of biological activity leads to draw-down of atmospheric carbon dioxide. But what is the fate of this carbon? Sequestration of anthropogenic CO_2 by deliberate fertilization of the ocean can only work if the organic carbon so produced can be exported to the deep sea, otherwise carbon dioxide will just 'burp' back into the atmosphere as the algae decompose above the permanent thermocline. Figure 2.13 shows that the flux of organic carbon sinking to the deep ocean increased following iron fertilization in the equatorial Pacific (and in fact the iron-stimulated biological community showed very high ratios of carbon export relative to carbon uptake), but no increase in the export of organic carbon following iron fertilization was observed in the Southern Ocean. This may be because waters are colder here, so the biological community is slower to respond to stimulation. Work is continuing in this field, but it seems clear that the effect of iron fertilization on carbon removal to the deep ocean is quite varied. A further note of caution is that there are potentially negative and deleterious ecosystem responses to iron addition that may exacerbate current global environmental problems; for example, increased organic matter production and subsequent decomposition may lead to a decline in the oxygen level of the water column, resulting in high mortalities across the biota.

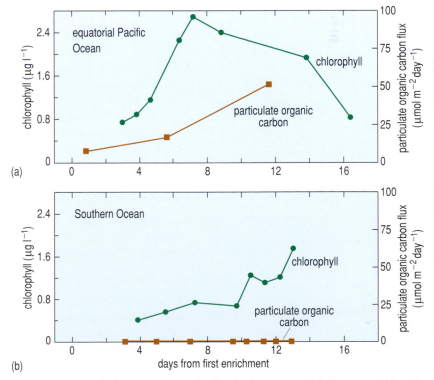

Figure 2.13   Biological response to iron fertilization in an area of (a) the equatorial Pacific Ocean and (b) the Southern Ocean. Also shown is the flux of particulate organic carbon sinking out of the surface ocean into the deep sea over the period of the experiment.

The biolimiting/biointermediate distinction is somewhat artificial in another sense, namely that the form of the profile of a dissolved constituent depends not only on the absolute amounts used by organisms but on the amounts used *relative to the total amount available in seawater*. It is clear, for instance, that much more carbon than phosphate is used in biological production; but there is also vastly more carbon than phosphate in solution and so it never becomes exhausted in surface waters.

The carbon dissolved in seawater is nearly all inorganic, mainly as the hydrogen carbonate (also know as bicarbonate) ion (HCO_3^-). Dissolved carbon is supplied directly to the ocean by rivers (mainly as dissolved hydrogen carbonate), by direct solution of CO_2 gas from the atmosphere, and by the respiration of marine organisms, which releases CO_2 into solution. We shall look at the carbonate system again in Chapter 3. The important point to recognize here is that the figure for carbon in Table 2.1 represents inorganic forms. The amount of carbon in the dissolved organic matter discussed at the end of Section 2.2.1 is very small by comparison.

Redfield ratios

The average molar ratio of carbon to the two principal nutrient elements in organic matter — nitrogen and phosphorus — is close to $106:16:1$ ($C:N:P$), and this is the basic **Redfield ratio**. The consumption, decomposition and recycling of organic particles as they sink through the water column results in progressive extraction of the nutrients (nitrate and phosphate) and an increase in carbon : nutrient ratios (i.e. $C:N$ and $C:P$ increase), and the particles become more refractory. Particles with high settling velocities have a much greater chance of reaching the sea-bed without significant decomposition, and $C:N$ ratios of $6:1$ or $7:1$ (i.e. with virtually unchanged Redfield ratios) are quite commonly found in the larger particles recovered in deep sediment traps. Once it reaches the sea-bed, this comparatively fresh material is rapidly decomposed by benthic animals and bacteria.

We have seen in Section 2.2 that the amount of particulate organic matter in seawater decreases with depth, and this must include the total population of bacteria. However, there is some evidence (Figure 2.14) that the *proportion* of bacteria in free suspension, relative to those colonizing larger particles, may actually increase with depth. Below about 2 km, bacteria contribute very little to the total biomass of rapidly sinking particles; observation and experiment suggest that such particles may be rather poor habitats for bacteria. This would help explain why a significant fraction of sinking particles reaches the sediments with a nutrient 'load' more or less intact.

Redfield ratios can also be worked out for the main skeleton-forming elements, calcium and silicon. In particulate matter dominated by siliceous organisms, for instance, the ratio $C:Si:N:P$ is about $106:40:16:1$.

2.2.3 PARTICULATE ORGANIC MATTER AND THE SCAVENGING CYCLE

The **adsorption** of metal ions or ionic complexes (whether in true solution or in colloidal suspension, e.g. oxides of iron or manganese) onto particle surfaces is a potent mechanism for removing trace elements from the seawater solution. Adsorption results from the mutual attraction between the charges on the ions and suitable bonding sites on the surfaces of particles, mainly the surfaces of bacteria. Bacteria make up nearly half of the total amount of particulate organic matter; they are thus very abundant, and their small size means that their total surface area is enormous. This is because the

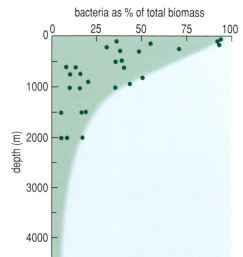

Figure 2.14   Estimates of the percentage (by weight) of the biomass of rapidly sinking organic particles contributed by bacteria. Edge of green area represents inferred upper limits. Below about 1–2 km depth, most bacteria are probably in free suspension.

ratio of surface area to volume increases with decreasing particle size: the ratio is about 3×10^3 for a particle of 1 mm radius, but 3×10^6 for a particle of 1 μm radius. For particles less than 5 μm in diameter, the total surface area amounts to $10 \, m^2$ per gram. It has been estimated that in the upper ocean something like 90% of the surface area of suspended particles is on living bacteria; and that the total surface area of freely suspended bacteria is two orders of magnitude greater than that of inorganic particles such as silt and clay.

Elements adsorbed onto small suspended particles are removed from the water column as large particles capture the small ones and carry them downwards — a process called **scavenging** (this term is sometimes used to encompass both parts of the process: adsorption and capture). It is important to stress that adsorption and scavenging are passive processes so far as bacteria are concerned. They act merely as an agent of transport, they do not make use of the elements they adsorb from solution. Nor are the trace elements adsorbed onto particle surfaces generally in great demand for the metabolic requirements of other marine organisms; although they will be ingested along with organic particles by filter-feeding organisms, they are soon excreted again, to become available for further scavenging.

Adsorption and scavenging is an extremely efficient way of removing some trace elements from the water column. The mean lifetime of small particles before they become part of larger particles in seawater is about 7.5 years. The mean residence times for **scavenged elements** are rather longer, because adsorption is not a one-way process: there are adsorption–desorption equilibria between ions in solution and those attached to particles, so scavenging cannot be 100% efficient. All the same, the residence time for scavenged elements is invariably less than 10^3 years, and for many it is less than 100 years.

In some cases, removal from the water column can be extremely rapid. Following the Chernobyl nuclear accident in 1986, for example, sediment traps in a number of places around Europe recorded dramatic increases in concentrations of radionuclides sinking to the sea-bed, within only weeks of the explosion. As soon as the fall-out reached the sea, surface-active nuclides (such as ^{137}Cs, ^{95}Nb, ^{95}Zr, ^{144}Ce, ^{103}Ru) were adsorbed onto particles which were in turn packaged into faecal pellets that sank rapidly to the sea-bed, perhaps more rapidly than would normally have been the case, because the accident more or less coincided with the spring plankton blooms.

What does all this tell us about the extent to which scavenged elements are mixed in the oceans?

As the mean oceanic mixing time is about 500 years, and residence times of most scavenged elements are much less than this, these elements cannot be uniformly mixed throughout the oceans (cf. Question 2.2). It follows that their distribution must to a very great extent reflect the influence of their sources either at the boundaries of the ocean (i.e., estuaries, the sea-bed) or within the water column.

For example, there is more dissolved aluminium in Atlantic than in Pacific surface waters. This is partly because more river water enters the Atlantic relative to its size (the ratio of the ocean area to land area drained by rivers is about six times smaller for the Atlantic than for the Pacific); and partly because a major source is aeolian (wind-blown) dust, especially in low

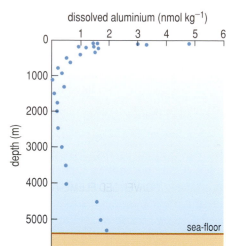

Figure 2.15   Profile for dissolved aluminium in the central North Pacific (28° 15′ N, 155° 07′ W). The increase in concentration at the bottom of the profile may be due to re-solution in deep water and/or to diffusion from sediment pore waters (see Chapter 5). (nmol = nanomole = 10^{-9} mole, and nmol $l^{-1} \approx$ nmol kg^{-1}.)

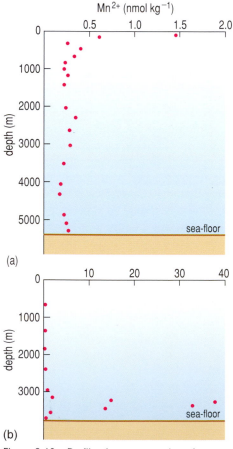

(a)

(b)

Figure 2.16   Profiles for concentration of dissolved manganese, for use with Question 2.5. Both are from stations in the North Atlantic: (a) 19° N, 23° W (off north-western Africa); (b) above the Mid-Atlantic Ridge at 26° N. Note the different horizontal scales.

latitudes, and the Atlantic is well supplied with dust from the Sahara, carried on the North-East Trade Winds. The aluminium is in clay minerals and particulate hydrated oxides ($Al_2O_3nH_2O$). Some of it goes into solution, but in surface waters of the oceans it is soon adsorbed and scavenged by sinking particles, so that its concentration decreases with depth (Figure 2.15).

The behaviour of manganese is generally similar to that of aluminium, but is complicated by the fact that manganese is more soluble as Mn(II) (Mn^{2+}) than as Mn(IV) (Mn^{4+}), and it is not always easy to distinguish between dissolved and particulate forms (cf. Note 5 to Table 2.1). A good deal of manganese is supplied to the oceans by rivers, so concentrations in surface waters tend to decrease away from land. In the open oceans, aeolian dust supplies manganese to surface waters, and in some regions there is another source of manganese, as you will see in Question 2.5.

QUESTION 2.5   Figure 2.16 shows profiles for dissolved manganese at two different stations in the Atlantic Ocean.

(a)   Bearing in mind the location and the shape of the profiles, can you suggest what is the likely principal source of manganese at each station?

(b)   According to the concentration scales, which source is the more important in quantitative terms?

The maximum in Figure 2.16(b) does not show up in Figure 2.16(a), either because advection of hydrothermal plume waters away from the ridge is in the wrong direction; or because the manganese is scavenged from the plume before it is advected into the area of profile (a); or both. In fact, anomalous concentrations of manganese due to hydrothermal inputs can be detected for hundreds of kilometres from the source (although they are eventually eliminated by the combined effects of mixing and particle (bacterial) scavenging).

The form of the profile in Figure 2.16(a) needs further explanation, because at first sight it seems to be a paradox. If aeolian dust is the main source of manganese to the surface ocean, then the manganese must be in particulate form, i.e. as oxidized and insoluble Mn(IV). So, how can concentrations of *dissolved* manganese be highest at the surface? There is evidence that manganese dissolves readily from dust in surface seawater, and it may be that complex photochemical reactions reduce insoluble Mn(IV) to soluble Mn(II). By contrast, manganese is supplied to the deep oceans by hydrothermal solutions (Figure 2.16(b)), mainly as the more soluble Mn(II). In both cases, the manganese is removed from solution in two ways: partly by direct adsorption and scavenging as Mn(II), followed by oxidation to Mn(IV) on particle surfaces; and partly by direct oxidation to particulate Mn(IV) in colloidal form, which is then scavenged by larger particles. In either case, particulate Mn(IV) can only be reduced to soluble Mn(II) again if the particles encounter an oxygen-deficient environment (see Section 2.5.3).

Because the distributions of scavenged elements reflect the influence of sources, they can be used as tracers to help map the movements of **water masses** through the oceans. Although scavenging is a passive process, a great deal of it is biologically driven, particularly in surface waters. Good correlations have been found between levels of primary production in the photic zone, the downward flux of particulate organic matter, and the removal of scavenged elements from solution. The Chernobyl fall-out example cited earlier is a case in point.

2.2.4 A CLASSIFICATION OF THE ELEMENTS IN SEAWATER

Figure 2.17 summarizes in diagrammatic form the characteristic profiles for the three main groups of elements we have just described and discussed, and lists the elements in each group. The conservative constituents interact only weakly with the biological particle cycle and nearly all have residence times of 10^6 years or more. They include the major constituents discussed in Section 2.1 as well as several trace elements, and are sometimes referred to as **bio-unlimited** constituents. The recycled and scavenged groups have been described and discussed in Sections 2.2.2 and 2.2.3 respectively. Of

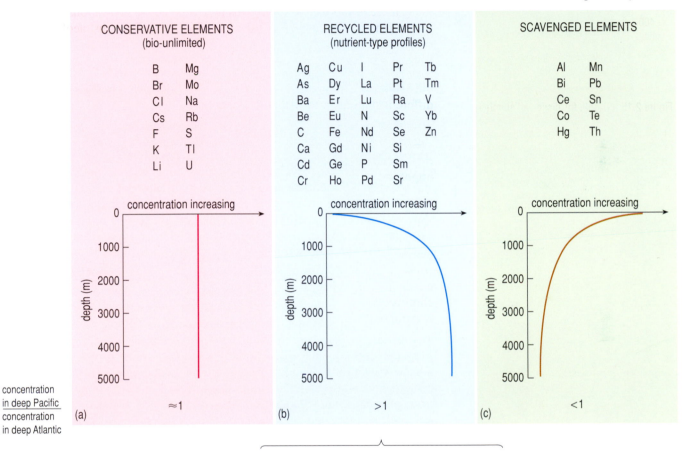

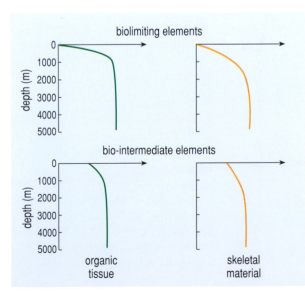

Figure 2.17 Summary classification of dissolved elements in seawater according to their profiles, with the elements listed in each category. Concentration ratios in deep ocean waters will be discussed in Section 2.4.
(a) Conservative or bio-unlimited elements.
(b) Recycled elements. This group has nutrient-type profiles and can be further subdivided into biolimiting or bio-intermediate elements according to criteria discussed in Section 2.2.2. Note that elements associated with organic tissue are recycled more rapidly than those associated with skeletal material.
(c) Scavenged elements.

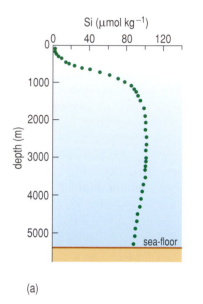

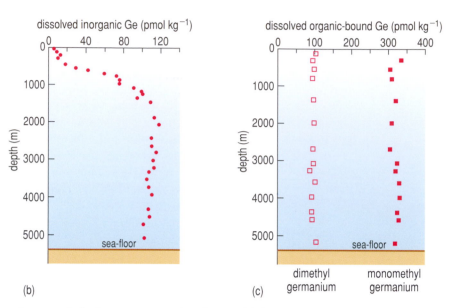

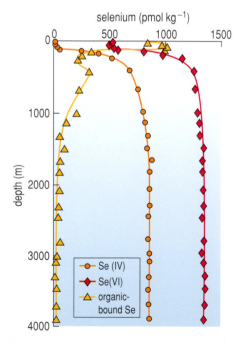

Figure 2.18   Profiles of (a) dissolved silica, (b) dissolved inorganic germanium (in the form of Ge(OH)$_4$), and (c) dissolved organic-bound germanium (monomethyl germanium and dimethyl germanium) for a site in the north-west Pacific.

the recycled elements, note that those which constitute the soft tissues (e.g. nitrogen, phosphorus) reach their maximum concentration at depths shallower than those that constitute the skeletal material (e.g. silica). This is because soft tissues are recycled more rapidly than skeletal material. Moreover, data for the different elements in these groups in Figure 2.17 show that some fit the patterns better than others, and new information could change the position of such elements in the classification scheme.

It is important also to stress that just because an element exhibits a typical recycled profile, it does not mean that the element cannot be scavenged from solution as well. This applies to (amongst others) nickel (Ni), vanadium (V), copper (Cu), zinc (Zn) and iron (Fe); conversely, a scavenged profile does not preclude involvement in the metabolism of organisms (e.g. manganese (Mn), cobalt (Co), and possibly aluminium (Al)) — the profiles merely indicate which of the two behaviour patterns is dominant. The behaviour of an element can also be linked to its **speciation**. For example, inorganic forms of germanium behave like silicon and are incorporated into siliceous skeletons, so inorganic germanium exhibits a recycled profile (Figure 2.18). By contrast, organic-bound (methylated) forms of germanium are biologically inert and behave conservatively in seawater.

QUESTION 2.6   Figure 2.19 shows profiles of three different forms of dissolved selenium (Se) in the Pacific. How would you classify (a) each form, and (b) all forms of dissolved Se combined, according to the scheme shown in Figure 2.17? (c) Which form of Se is preferentially utilized by marine organisms, and how can you tell?

An interesting phenomenon is sometimes called *uptake by analogy*. A number of elements in the recycled group of Figure 2.17 may be there because of their chemical similarity to more abundant vertical neighbours in the Periodic Table (Appendix 1), rather than because they have an essential biological role (indeed, some of them may even be toxic). Such pairs include arsenic (As) and phosphorus (P), silver (Ag) and copper (Cu), and palladium (Pd) and nickel (Ni). Until quite recently, it was thought that cadmium (Cd) and zinc (Zn) were another pair, but it is now known that Cd, like Zn, is required in metalloenzymes that are utilized by biological organisms for the acquisition of inorganic carbon.

Figure 2.19   Profiles of concentration of dissolved Se in the Pacific.

QUESTION 2.7 Whereabouts in the Periodic Table would you expect to find (a) strontium in relation to calcium, (b) germanium in relation to silicon?

The scavenging cycle and sediment fluxes

As well as through the deployment of sediment traps, rates of deposition and accumulation of sediment can be measured by dating successive layers using natural radio-isotopes such as carbon-14 (^{14}C) and the daughter products of uranium and thorium decay. Sediment fluxes can also be estimated by making use of the contrasting behaviour of different radio-isotopes with respect to the scavenging cycle. For example, one of the parent–daughter isotope pairs in the uranium–thorium decay series is ^{238}U–^{234}Th. Uranium behaves conservatively in seawater; but thorium falls in the scavenged category (Figure 2.17). By measuring the concentration of dissolved uranium in the water column at a particular depth, we can calculate the expected concentration of dissolved ^{234}Th at that depth from the known decay constant. The actual concentration of dissolved ^{234}Th will be less than the calculated value by an amount that depends on how much has been scavenged from solution, and this can in turn be related to the vertical particle flux at that depth: the greater the depletion of dissolved ^{234}Th in the water column, the more has been scavenged because of the higher particle flux.

Another way of estimating sediment fluxes is to quantify the movements of dissolved constituents as they travel through the oceans from sources to sinks. This is the subject of the next Section.

2.3 VERTICAL MOVEMENT OF DISSOLVED CONSTITUENTS: THE TWO-BOX MODEL

Marine chemists find it convenient to model the oceans as a layered series of well-mixed reservoirs. Here we shall confine ourselves to only two such reservoirs, but this will nonetheless enable us to quantify the basic processes occurring in the oceans and to estimate, to a first approximation, the rates at which different constituents move through the system. The two layers into which we can simplify the oceans as a whole are a thin upper layer of warm water, and a much larger cold reservoir beneath it.

What is the boundary separating these two reservoirs?

It is the base of the **mixed surface layer**, between 100 and 200 m depth on average. This is the bottom of the upper 'box'. Below it lies the permanent thermocline extending down to 500–1000 m over the most of the world's oceans, and below that are the intermediate and deep-water masses. These all make up the lower 'box' which is thus some 20 times bigger than the upper 'box'. This simple two-box model can be more easily used if we make further simplifying assumptions:

1   Dissolved constituents are added to seawater by rivers only. Sources such as aeolian, volcanic or hydrothermal inputs, or the inflow of groundwater at continental margins, are ignored.

2   The only way dissolved constituents are removed from the ocean is by organic (biogenic) particles sinking to the sea-floor.

3   The ocean is a steady state: rates of input and loss of any dissolved constituents, both in the ocean as a whole, and between the surface and deep reservoirs, have remained constant for long periods, so that concentrations at any point do not change with time.

It follows from this set of assumptions that dissolved material *added* by rivers (the source) to the sea must (after being mixed and recycled within the oceans) be *removed* at the same rate, by preservation in sediments accumulating on the bottom (the sink).

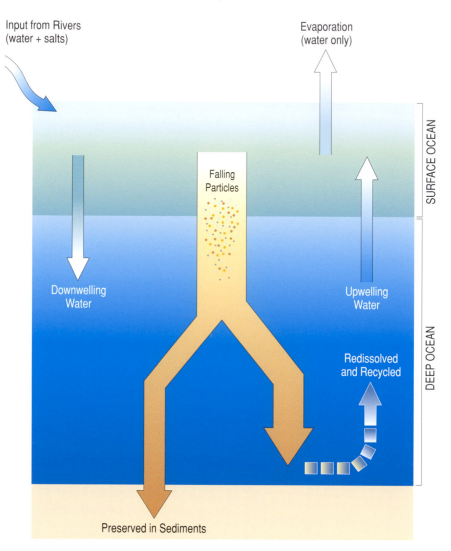

Figure 2.20   Simple two-box model for the oceans. The numbered notes below explain the various fluxes and their relationships:

1   The net Input from Rivers consists of dissolved constituents only, because the hydrological cycle ensures that an equal amount of water is removed by Evaporation.

2   The material supplied by rivers in solution represents new material that must be removed from the system if a steady state is to be maintained. So the Input from Rivers is balanced by material Preserved in Sediments.

3   Material from the surface ocean is lost to the deep ocean in Downwelling Water (partly simple downward mixing, but mainly the formation of deep-water masses) and in Falling Particles.

4   The volume of water in both reservoirs must remain unchanged, however, so the Downwelling Water must be balanced by an equal amount of Upwelling Water (this is mainly simple upward mixing, but includes localized upwelling).

5   When the Falling Particles reach the bottom of the deep reservoir, a proportion equivalent to the Input from Rivers is Preserved in Sediments, but the remainder will be Redissolved and Recycled, to be carried up to the surface ocean in due course by the Upwelling Water.

6   *Very important:* Biologically active elements have different concentrations in surface and deep oceans (e.g. Figure 2.10), so very different amounts of these elements will be transported by Downwelling and Upwelling Waters.

QUESTION 2.8   It also follows from those assumptions that the two-box model can be applied only to dissolved constituents in the recycled category of Figure 2.17, but not to others. Why is that?

We shall now set up and explain the **two-box model** in diagrammatic form, and then use it to examine the behaviour of a biolimiting nutrient: phosphate.

Study Figure 2.20 and read its caption carefully to make sure that you understand the balance between the different fluxes, before we apply the model quantitatively in the following exercise.

2.3.1   THE TWO-BOX MODEL AND PHOSPHATE

Figure 2.21 is a version of Figure 2.20 that contains spaces for you to insert the values of the fluxes and balances, after following through the simple calculations set out below. The completed version of Figure 2.21 is provided at the back of this Volume (Figure A3).

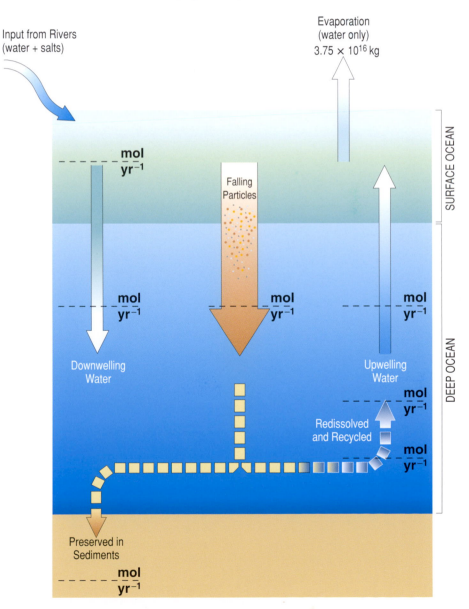

Figure 2.21   Two-box model for use with the phosphate exercise. For the completed calculation, see Figure A3, immediately before the answer to Question 2.9.

Throughout this exercise, we shall use molar concentrations. Concentrations in mol l^{-1} are virtually the same as concentrations in mol kg^{-1} (because 1 litre of water weighs nearly 1 kilogram and 1 litre of seawater weighs just over 1 kilogram).

To calculate the Input from Rivers, we need to know the annual flux of water from the world's rivers into the oceans, and the concentration of phosphate in river water. Some 3.75×10^{16} kg of water are supplied annually to the oceans by rivers. The average concentration of phosphate is about 0.5×10^{-6} mol PO$_4^{3-}$ kg^{-1}, so we can work out the annual Input from Rivers (in moles) and insert the result in Figure 2.21:

Input from Rivers = $(3.75 \times 10^{16}$ kg yr$^{-1}) \times (0.5 \times 10^{-6}$ mol PO$_4^{3-}$ kg$^{-1})$
$= 18.75 \times 10^{9}$ mol yr^{-1} PO$_4^{3-}$.

(Note that the Evaporation figure is the same as the annual flux of river water, and we have inserted this in Figure 2.21.)

The Input from Rivers must be the *same* as the amount Preserved in Sediments, for the steady-state composition of the ocean to be maintained. So, you can make another entry in Figure 2.21: 18.75×10^{9} mol yr^{-1} PO$_4^{3-}$ Preserved in Sediments.

Oceanographers have used ^{14}C age-dating techniques to work out that the upwelling flux of seawater is about 20 times the river flux. Upwelling and downwelling fluxes must be equal, and so you can easily work out the amount of *water* that is being exchanged between the surface and deep oceans each year.

What is it?

As the river flux is about 3.75×10^{16} kg yr^{-1}, then both the upwelling and downwelling flux must be in the order of 7.5×10^{17} kg yr^{-1}.

What is the principal feature of a biolimiting recycled constituent such as phosphate with respect to its concentration in the surface and deep oceans?

Concentrations are very much less in the surface than the deep oceans, and for phosphate these concentrations are in the order of 0.1×10^{-6} mol PO$_4^{3-}$ kg^{-1} and 2.5×10^{-6} mol PO$_4^{3-}$ kg^{-1}, respectively (Figure 2.10).

You now have the water flux for Downwelling and Upwelling Waters, and the phosphate concentrations in the surface and deep oceans, so you can work out respective phosphate fluxes for Figure 2.21:

PO$_4^{3-}$ in Downwelling Water = $(7.5 \times 10^{17}$ kg yr$^{-1}) \times (0.1 \times 10^{-6}$ mol PO$_4^{3-}$ kg$^{-1}) = 75 \times 10^{9}$ mol yr^{-1} PO$_4^{3-}$.

PO$_4^{3-}$ in Upwelling Water = $(7.5 \times 10^{17}$ kg yr$^{-1}) \times (2.5 \times 10^{-6}$ mol PO$_4^{3-}$ kg$^{-1}) = 1875 \times 10^{9}$ mol yr^{-1} PO$_4^{3-}$.

If the concentration in the surface ocean is to be maintained at its low level, then the huge difference between the phosphate fluxes for Upwelling Water and Downwelling Water must be balanced.

By what?

Most of the phosphate that is added from rivers and that mixes up from the deep ocean is fixed by organisms in surface waters. These organisms subsequently die and sink, carrying the phosphate with them, so they transport the balance of the phosphate that is not carried to the deep ocean by Downwelling Water.

The amount of phosphate carried down by Falling Particles must therefore be given by:

PO_4^{3-} in Particles = PO_4^{3-} in Upwelling + PO_4^{3-} in Rivers − PO_4^{3-} in Downwelling

= $(1875 \times 10^9 \, mol \, yr^{-1} \, PO_4^{3-}) + (18.75 \times 10^9 \, mol \, yr^{-1} \, PO_4^{3-}) - (75 \times 10^9 \, mol \, yr^{-1} \, PO_4^{3-}) = 1818.75 \times 10^9 \, mol \, yr^{-1} \, PO_4^{3-}$.

Now all that remains is to work out the amount of phosphate that is Redissolved and Recycled in the deep ocean.

What will that amount to?

It will be the difference between the amount of phosphate carried down by Falling Particles, and the amount that is Preserved in Sediments:

$(1818.75 \times 10^9 \, mol \, yr^{-1} \, PO_4^{3-}) - (18.75 \times 10^9 \, mol \, yr^{-1} \, PO_4^{3-}) = 1800 \times 10^9 \, mol \, yr^{-1} \, PO_4^{3-}$.

As a check: PO_4^{3-} in Upwelling = PO_4^{3-} Redissolved and Recycled + PO_4^{3-} in Downwelling:

$(1875 \times 10^9 \, mol \, yr^{-1} \, PO_4^{3-}) = (1800 \times 10^9 \, mol \, yr^{-1} \, PO_4^{3-}) + (75 \times 10^9 \, mol \, yr^{-1} \, PO_4^{3-})$

To make useful comparisons of the results of two-box model calculations for different constituents, we need to work out:

1 What percentage of the constituent entering the surface ocean (from both Rivers and Upwelling) (a) goes into Falling Particles, (b) is Preserved in Sediments.

2 What percentage of the constituent in Falling Particles is Preserved in Sediments.

QUESTION 2.9 Now do these additional calculations (1 and 2) for phosphate.

So far as phosphate is concerned, then, nearly all of the phosphate that enters the surface ocean is removed in Falling Particles. However, only about 1% of the Falling Particles survives to be Preserved in Sediments — 99% is Redissolved and Recycled. Only about 1% of the phosphate entering the surface ocean (from *both* Rivers *and* Upwelling) is Preserved in Sediments. Put another way, each mole of phosphate goes through an average of about 100 cycles within the ocean before it is incorporated into sediments on the ocean floor.

The two-box model is a greatly simplified approximation of the real ocean. However, the *basic principles* of this model are valid, and the results are realistic enough to bring out the more important differences in the behaviours of recycled constituents, *provided you do not quote those results to umpteen significant figures*.

2.3.2 ESTIMATING RESIDENCE TIMES

One of the essential pieces of information for two-box model calculations is the figure for Input from Rivers. From the definition given earlier:

$$\text{residence time } (\tau) = \frac{\text{total mass of substance dissolved in the oceans}}{\text{rate of supply (or removal) of the substance}}$$

We can use the figure for Input from Rivers with data from Table 2.1 to estimate residence times.

Caution: The residence time calculation is simple, but you must be careful to use the right numbers. In Section 2.3.1, concentrations and quantities are in molar terms, in Table 2.1 they are in terms of weight. The relationship is:

> molar quantity × (relative atomic (or molecular) mass) = weight quantity (in grams)

The relative atomic mass of phosphorus is 31. On average, 18.75×10^9 mol of PO_4^{3-} enter the oceans from rivers each year (Input from Rivers, Figures 2.20 and A3). This is equivalent to $18.75 \times 10^9 \times 31 \approx 580 \times 10^9$ g of P, which is 580×10^3 tonnes annually (remember that 1 mol of phosphate (PO_4^{3-}) contains 1 mol of phosphorus (P)). The total mass of phosphorus in the oceans from Table 2.1 is 7.93×10^{10} tonnes, and so the residence time must be:

$$\frac{7.93 \times 10^{10} \text{ tonnes}}{580 \times 10^3 \text{ tonnes yr}^{-1}} \approx 1.4 \times 10^5, \text{ or } 140\,000 \text{ years}$$

(assuming that all phosphorus is in the form of phosphate).

If phosphorus were supplied to the oceans by hydrothermal processes as well as by rivers, would the residence time be longer or shorter than the one we have just calculated?

It would be shorter, because the rate of supply would be greater. However, residence time calculations can at best be only approximate estimates, because of uncertainties in determining rates of supply of dissolved constituents on the one hand, and the total amount of a particular constituent in the oceans on the other.

2.4 LATERAL VARIATIONS OF DISSOLVED CONSTITUENTS IN THE DEEP OCEANS

The two-box model is very useful for estimating vertical fluxes but it ignores the fact that ocean waters are continually moving, and that a profile of (say) phosphate is partly a result of the contribution of different water masses flowing at various depths.

Figure 2.22 shows how the concentration of phosphate varies in deep water. From a minimum in the North Atlantic, it increases southwards and continues to increase round southern Africa, eastwards and northwards in the Pacific. Table 2.2 summarizes concentration ratios in the Atlantic and the Pacific for four elements used in biological processes. Note that the ratio uses the *difference* between deep and surface concentrations in each

Table 2.2 Average ratios for biologically important elements in the deep parts of the Atlantic and the Pacific (C = concentration)*.

Element	$\dfrac{(C_{deep} - C_{surface}) \text{Pacific}}{(C_{deep} - C_{surface}) \text{Atlantic}}$
Nitrogen (as NO_3^-)	2
Phosphorus (as PO_4^{3-})	2
Carbon	3
Silica	5

*Note that the geographic distribution of deep-ocean enrichment for each element is qualitatively similar to that for phosphorus (Figure 2.22); and on a global scale, surface concentrations for a given element are broadly similar in all oceans.

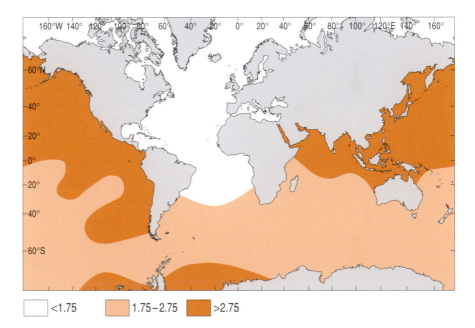

Figure 2.22 Distribution of the concentration of dissolved phosphate in seawater at 2000 m depth. Units are 10^{-6} mol l^{-1}.

	<1.75		1.75–2.75		>2.75

ocean, so Table 2.2 is a measure of the *lateral* enrichment of each element in deep Pacific water compared with deep Atlantic water.

QUESTION 2.10   Is there any obvious difference between the ratios for hard-part and soft-part constituents in Table 2.2? Why might this be?

Bearing in mind that the ratios in Table 2.2 express an enrichment in the *deep* oceans (which is also illustrated in Figure 2.22), we need to look for a mechanism to explain why the deep Pacific is so much richer in nutrients than the deep Atlantic.

This mechanism is summarized in Figure 2.23. Nutrient-poor **North Atlantic Deep Water** (NADW) flows southward in the western Atlantic and is steadily enriched with nutrients derived from the rain of particulate organic matter sinking from the surface and being redissolved in the deep ocean. It then flows round southern Africa and into the Indian and Pacific Oceans. In the southern Atlantic, it is joined by **Antarctic Bottom Water** (AABW), which is more nutrient-rich than NADW to start with, because of upwelling along the **Antarctic Divergence**, and decomposition of the resulting 'rain' of particulate organic matter. Although some AABW

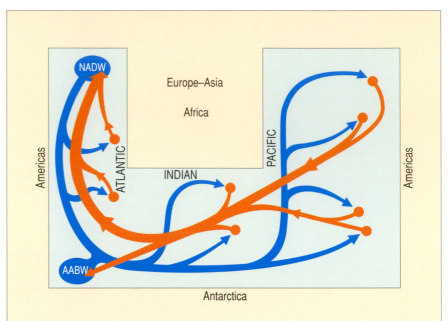

(a)

(b)

Figure 2.23 (a)　Schematic map of the overall
effect of the thermohaline circulation (the
'thermohaline conveyor'). The orange part of
the conveyor represents the net transport of
warm water in the uppermost 1000 m or so, the
blue part the net transport of cold water below
the permanent thermocline. Large ellipses
represent sources of North Atlantic Deep Water
(NADW) and Antarctic Bottom Water (AABW);
small orange circles represent areas of localized
upwelling.
(b)　Schematic cross-section from the North
Atlantic to the North Pacific, showing major
advective flow patterns (thin lines) and the rain
of particles (thick arrows). AABW is not shown
in this picture.

flows northwards in the Atlantic, much of it flows eastwards into the Indian
and Pacific Oceans, along with NADW, and nutrient enrichment
by decomposition of sinking particulate organic matter continues. Thus,
the oldest and most nutrient-rich waters are in the deep Pacific Ocean
(Table 2.2).

Deep water is returned to the surface mainly by the broad, slow and rather
diffuse upwelling that goes on continuously throughout the world ocean.
The localized coastal and equatorial upwelling areas schematically identified
in Figure 2.23(a) also play a role, but much of the upwelled water in these
regions comes from above the permanent thermocline, *not* from the deep
ocean.

All of this has a bearing on the significance of a mean oceanic mixing time,
and suggests that the deep water in some parts of the oceans is much older
than the average value of about 500 years, i.e. it has been away from the
surface for much longer than 500 years. Radiocarbon measurements of deep
waters in parts of the northern Indian and Pacific Oceans yield values in
excess of 1000 years, while in other regions the exchange between deep and
surface water takes much *less* than 500 years. As is so often the case, the
average thus conceals considerable variations.

QUESTION 2.11 In general, the older the deep water, the richer it will be in elements with recycled (nutrient-type) profiles (cf. Figure 2.17). That explains the concentration ratios of >1 for the elements in Table 2.2. Can you explain why concentration ratios for scavenged elements are less than 1?

Figure 2.24 summarizes the overall distribution of the principal biologically important dissolved constituents of seawater in different parts of the ocean. The diagonal band shows that the molar ratio of N : P remains constant at 16 : 1 throughout the range of Figure 2.24 (i.e. it is the same as the molar N : P ratio in organic tissue, the Redfield ratio, Section 2.2.2). Thus, when all the dissolved nitrate in surface waters has been used up, so has all, or nearly all, the dissolved phosphate — and *vice versa*. The occurrence of nitrate and phosphate in seawater in the same ratio that organisms require is unlikely to be just a coincidence. Instead, it has probably been brought about over time by the actions of nitrogen-fixing phytoplankton, such as the free-living cyanobacterium *Trichodesmium*, which do better in competition with other phytoplankton when phosphate is present but nitrate is absent, but are outcompeted otherwise. Nitrogen fixers obtain their nitrogen from dinitrogen (N_2, molecular nitrogen) rather than nitrate, but their decomposition raises levels of nitrate in the same way as in the case of other phytoplankton. Abundant growth of nitrogen-fixing phytoplankton, inevitably followed a short time later by their decay, thereby raises NO_3^- relative to PO_4^{3-}. This mechanism thus releases extra nitrate when nitrate runs out but when there is still some phosphate available, creating a dynamic link between the levels of NO_3^- and PO_4^{3-}. Growth of the nitrogen-fixing phytoplankton will continue until phosphate runs out.

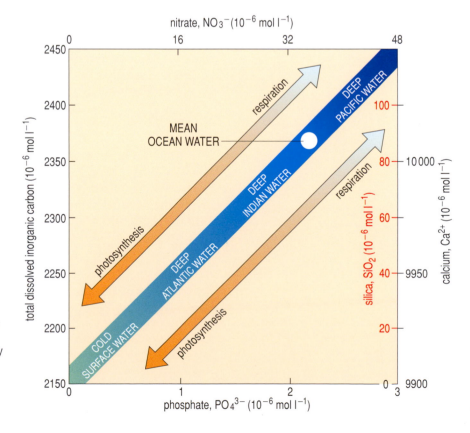

Figure 2.24 Idealized diagram to illustrate changes in nutrient and skeleton-building elements in the oceans as a result of biological activity. Conditions at the surface water end (lower left) of the diagonal blue band are achieved when the biolimiting nutrients have been exhausted as a result of photosynthetic activity by phytoplankton. The other end is determined by the degree of enrichment in the deep sea from decomposition of organic matter as a consequence of respiration by marine organisms. Intermediate values are produced by mixing these end-members in all proportions. The axis values for calcium and silica correspond to water where hard parts are formed, and are very generalized.

2.5 THE AIR–SEA INTERFACE

The transfer of heat and water across the ocean–atmosphere boundary is of great importance in controlling Earth's climate. Transfers of gases, liquids and solids across the interface are important components of the biogeochemical cycles that occur within the oceans; for example, it is estimated that the atmospheric supply of nitrogen to the oceans is comparable to that from rivers, while the main external source of iron to the oceans is from atmospheric dust deposition. Transfers take place across a thin sea-surface boundary layer some 20–200 μm thick. Hydrophobic (water-repellent) organic molecules such as fats and oils, both natural and artificial, tend to accumulate at the surface of this layer. The layer itself is almost completely transparent to solar radiation, and just below it, in the uppermost few millimetres of the water column, live the **neuston**, which include phytoplankton specially adapted to survive and utilize high intensities of ultraviolet light. Zooplankton in the neuston are frequently coloured in shades of blue.

2.5.1 TRANSFER OF GASES

The exchange of gases across the surface boundary layer is driven by the concentration difference between air and surface seawater, transport by **molecular diffusion**, and, where wind speed is sufficient, turbulent motion. Transfer of gases is a two-way process, even at saturation, when the concentration of gas in the water is in equilibrium with the concentration of the gas in the atmosphere, so that gas transfer is equal in both directions.

The basic equation describing the net flux, F, of a gas across an air–water boundary is:

$$F = K\Delta C \qquad\qquad (2.1)$$

where K is the transfer velocity, the rate at which transfer occurs, and ΔC is the concentration difference driving the gas flux across the interface (Figure 2.25): K is usually expressed in cm hr^{-1}, and if C is in ml l^{-1}, then F will be in ml cm^{-2} hr^{-1} — i.e. unit volume crossing unit area in unit time.

Would you expect values of K in Equation 2.1 ever to reach zero?

The form of Equation 2.1 is such that K is in effect a proportionality constant (it is often called the exchange coefficient). At saturation, where ΔC is effectively zero, there will still be exchange of gas through the boundary layer, but it will be the same in both directions. The *net* flux, F, will thus be zero. The only conditions under which K itself might approach zero are where a 'lid', such as a layer of ice or a thick oil slick, covers the surface.

In normal circumstances, values of K range from a few centimetres per hour to a few tens of centimetres per hour and increase with increasing wind speed (Figure 2.26). At low wind speed, surface waters are smooth and gas transfer occurs only by molecular diffusion. As wind speed increases, waves become common (surface waters are 'rough') and gas transfer due to vertical turbulence becomes more important than molecular diffusion. At high wind speeds (>12 m s^{-1}), bubble bursting further enhances gas transfer rates.

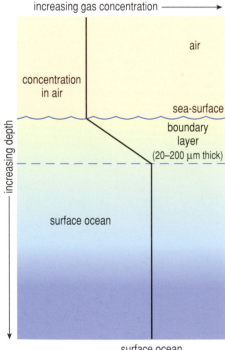

increasing gas concentration →

air

concentration in air

sea-surface

boundary layer (20–200 μm thick)

surface ocean

increasing depth

surface ocean concentration

Figure 2.25 Relative gas concentrations in the air, boundary layer, and underlying surface water (not to scale). In this case, the concentration of the gas is greater in seawater than in air, so there will be a net flux of the gas from seawater to air.

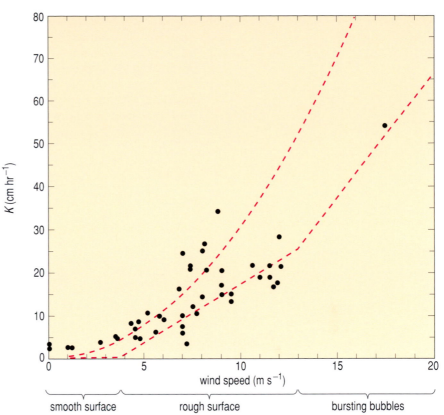

Figure 2.26   The effect of wind speed on transfer velocity. Data are obtained from field measurements (solid circles) and from models based on measurements made in wind tunnels (dashed lines). Note that the spread of the data increases with increasing wind speed: this is largely because winds tend to be gusty at high wind speeds, so wind speed must be averaged over the period for which K is measured.

Vertical turbulence due to the action of wind not only increases K but it also reduces the thickness of the boundary layer. Thus, the thickness of the boundary layer is not constant, and gas exchange is more rapid during storms than in calm weather. The solubility of all gases is greater in cold than in warm water: if the temperature of the water rises, then under equilibrium conditions there will be a net flux of gas from sea to air. In addition, gas solubility varies inversely with salinity — so the higher the salinity, the smaller the amount of any gas in solution.

Gases tend to fall into two categories:

1   Gases of low solubility and low chemical reactivity in water, such as nitrogen (N_2), oxygen (O_2), argon (Ar) and other noble gases.

2   Gases of high solubility and/or high chemical reactivity in water, such as sulphur dioxide (SO_2), nitrogen dioxide (NO_2) and ammonia (NH_3).

Major gases

The four most abundant atmospheric gases are nitrogen, oxygen, argon and carbon dioxide, in that order. To a first approximation, their concentrations in surface seawater are in equilibrium with their **partial pressure** (which is the same as their percentage by volume) in the atmosphere.

To which of categories 1 and 2 above do these four gases belong?

The first three are in category 1 (low solubility and reactivity). You might expect CO_2 to be in the second category, because it reacts with seawater to form hydrogen carbonate (bicarbonate) and other ions (see also Chapter 3). This increases its solubility, but it reacts more slowly than SO_2, etc., and it is generally grouped with the gases in category 1.

Figure 2.27 shows that oxygen is generally supersaturated in surface seawater. Some of the excess comes from oxygen released by photosynthesizing phytoplankton. Another cause of the supersaturation is air bubbles being carried by breaking waves down into the water column, where some of the air dissolves because of the increased hydrostatic pressure. The other atmospheric gases are also slightly supersaturated in surface waters, just as oxygen is.

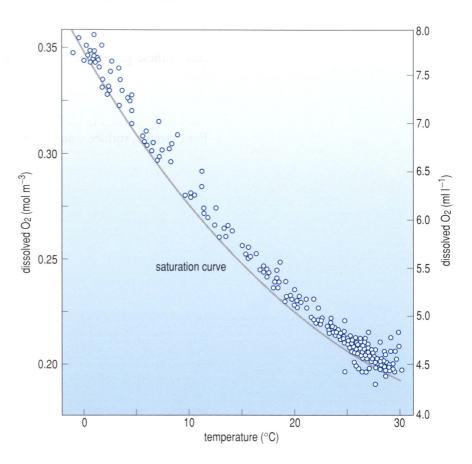

Figure 2.27   The saturation curve for oxygen (grey curve) and measured concentrations of dissolved oxygen in surface ocean waters (open blue circles).

QUESTION 2.12   Do the four main atmospheric gases behave conservatively or non-conservatively in seawater?

Although carbon dioxide is the most soluble of the major gases, its concentration in seawater as dissolved gas is very small. Nearly all the carbon dioxide in seawater is combined with water as the dissociation products of carbonic acid, HCO_3^- and CO_3^{2-}. The solubility relationships of CO_2 are controlled not only by equations such as Equation 2.1, but also by the chemical equilibria governing the reactions of the aqueous carbonate system (see Chapter 3). These reactions can help to 'mop up' the excess atmospheric CO_2 produced by human activities and thus act to moderate global warming.

Minor gases
These fall into both categories 1 and 2 above, and, like the major gases, some minor gases behave conservatively, others non-conservatively. It is not easy to quantify transfers of these gases across the air–sea interface, because of uncertainties inherent in determining both K (Figure 2.26) and ΔC in Equation 2.1. Uncertainties in ΔC are due principally to the

difficulty of measuring very low concentrations of minor gases in the marine environment (down to 1 part in 10^{12} or less).

For some minor gases, the net flux is from atmosphere to oceans, and this applies particularly to many pollutant gases, which include industrial additions to the natural SO_2 load, and other gases such as CFCs (chlorofluorocarbons) and PCBs (polychlorinated biphenyls).

Gases produced in ocean surface waters by planktonic organisms include methane (CH_4) and nitrous oxide (N_2O), which are both important long-lived greenhouse gases, and dimethyl sulphide (DMS) — (($CH_3)_2S$). The flux of these gases from sea to air is considerable.

Why should that be?

They are produced in surface waters and they are supersaturated there. For example, surface waters can have concentrations of DMS as high as $3 \times 10^{-6}\,g\,l^{-1}$ in regions of high biological production; whereas the concentration in equilibrium with normal atmospheric concentrations is three orders of magnitude less (c. $3 \times 10^{-9}\,g\,l^{-1}$). Production of DMS helps to regulate Earth's climate: it is oxidized in the atmosphere, forming sulphate aerosols which provide additional nuclei for water droplets and thus aid cloud formation. Clouds reflect solar radiation back to space, cooling the Earth's surface.

2.5.2   TRANSFER OF LIQUIDS AND SOLIDS

In this context, the liquid is rainwater and it contains particles, dissolved gases and salts. For air–sea transfer, the flux of material across the interface obviously depends primarily on the intensity of precipitation (rain, snow or hail), but it is also proportional to the **washout ratio**, w, which is defined as:

$$w = \frac{C'}{C} \tag{2.2}$$

where C' is the concentration of particles or gas in near-surface precipitation, and C is the concentration in the atmosphere.

Washout ratios can be determined experimentally, and they are useful in an environmental context because they are a measure of the ability of water droplets to remove particles or gas from the atmosphere. For particles, values of w range from 10^2 to between 10^5 and 10^6. For gases of low solubility and low chemical reactivity in water, w is much smaller, in the range 10^{-2} to 1; and for gases that are more soluble and/or chemically reactive, w ranges from 10^2 to 10^4, and exceptionally up to 10^5.

QUESTION 2.13   These values of w have some interesting implications.

(a)   Near the sea-surface, are (i) unreactive or (ii) reactive gases more or less concentrated in rain than in the atmosphere?

(b)   In general, which does rain 'scavenge' more efficiently from the atmosphere — particles or gases?

In the absence of rainfall, *dry deposition* of particles to the sea-surface will continue simply by direct fall-out from the atmosphere; the larger the particle, the greater its settling velocity and the more rapid the fall-out. To a very simple first approximation, for particles smaller than about

0.1 µm, such as occur in smoke, the transfer process is essentially diffusive, i.e. the particles behave rather like gas molecules.

The subsequent fate of elements delivered to the ocean in particulate form initially depends on the extent to which they are solubilized in seawater. Solubility is partly controlled by the nature of the particle (e.g., pollutant aerosols tend to be more soluble than dust-rich aerosols), the length of time the particle remains in the water column, and the chemical composition of the seawater: in particular, the presence of suspended particles and dissolved organic matter. Elements that are very soluble (> 90%) include sodium (Na) and bromine (Br); cadmium (Cd), zinc (Zn) and lead (Pb) are moderately soluble (20–90%). Elements such as aluminium (Al) and iron (Fe) are only slightly soluble (< 10%). Nevertheless, the atmospheric flux of Fe is important for regulating primary production in some parts of the ocean (Box 2.2).

The transfer of water and solids from *sea to air* by bubble bursting and aerosol formation (Figure 2.2) should not be forgotten. Aerosol droplets transport not only dissolved salts (ions), but also organic matter from surface layers; some of this is returned directly to the sea by rain or dry deposition, but some must fall on land — so there is a degree of exchange of organic carbon between oceans and continents.

2.5.3 OXYGEN IN SEAWATER

Marine animals and bacteria that consume and decompose organic matter require dissolved oxygen for respiration. The excess of abstraction over replenishment of oxygen reaches a maximum at around 1 km depth, where there is an **oxygen minimum layer**. Oxygen profiles generally mirror those for nutrients (Figure 2.28).

We can simplify the essentially reversible reactions that take place during formation of organic matter by photosynthesis, and its subsequent destruction by bacterial respiration (oxidation), in the following way:

If we represent organic matter by CH_2O (for fixed carbon) and NH_3 (for fixed nitrogen), then:

For carbon:

$$CO_2(g) + H_2O \rightleftharpoons CH_2O(s) + O_2(g) \tag{2.3}$$

For fixed nitrogen:

$$NO_3^-(aq) + H_2O + H^+(aq) \rightleftharpoons NH_3(s) + 2O_2(g) \tag{2.4}$$

These reactions move to the right during photosynthesis, to the left during respiration. Reaction 2.3 tells you that for every mole of carbon fixed by photosynthesis, one mole of molecular oxygen (O_2) is liberated; and reaction 2.4 shows that for every mole of nitrogen fixed, two moles of molecular oxygen are liberated. We can thus include oxygen in the Redfield ratio, which was introduced in Section 2.2 and which approximates to 106 : 16 : 1 (C : N : P).

QUESTION 2.14   Can you explain the shape of the profile for pH in Figure 2.28?

From the 1 : 1 molar relationship between carbon and oxygen in reaction 2.3, we can conclude that for every 106 moles of carbon (C) fixed in organic tissue by photosynthesis (or released from it as CO_2 by respiration), 106 moles of oxygen are liberated by photosynthesis (or consumed by respiration). From the 1 : 2 molar relationship between nitrogen and oxygen in reaction 2.4, we can further conclude that another 32 moles of oxygen are

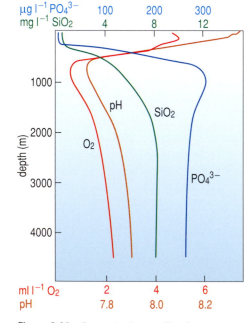

Figure 2.28   Concentration profiles from 24° 22′ N, 145° 33′ W (central sub-tropical Pacific), to show the contrasting shapes for dissolved nutrients and dissolved oxygen. Note that the maximum for silica is reached at greater depth than that for phosphate (cf. Figure 2.10 and also the profiles at the bottom of Figure 2.17). For pH profile, see Question 2.14.

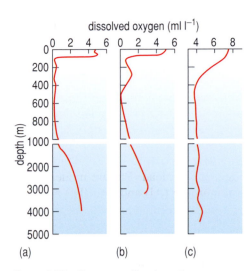

Figure 2.29 Oxygen profiles from three locations, to show the variable form of oxygen minimum layers. Note the change of scale at 1000 m depth. (a) Eastern tropical Pacific (11° 39′ N, 114° 15′ W). (b) Eastern tropical Pacific (6° 21′ N, 103° 42′ W). (c) Antarctic Polar Frontal Zone (50° 08′ S, 35° 49′ W). Sub-oxic conditions have developed in (a) and (b).

liberated by photosynthesis (or consumed by respiration) for every 16 moles of the fixed nitrogen that accompany the carbon in organic tissue. Phosphorus remains largely as phosphate during biological processes and does not participate in this cycle of oxidation and reduction to any significant extent. We can accordingly extend the Redfield ratio as follows:

$$\left.\begin{array}{l} 106 \text{ moles of carbon (C) are equivalent to} \\[4pt] 16 \text{ moles of nitrate } (NO_3^-) \text{ are equivalent to} \\[4pt] 1 \text{ mole of phosphate } (PO_4^{3-}) \text{ is equivalent to} \end{array}\right\} 138 \text{ moles of oxygen } (O_2)$$

Armed with this information and knowledge of how much oxygen and organic matter there is in seawater, we can go some way to explaining semi-quantitatively why there is a certain range of depths where the ocean can be quite strongly depleted in dissolved oxygen, i.e. why oxygen minimum layers form (Figure 2.29).

For example, an oxygen concentration in surface water of 6 ml l^{-1} approximates to $0.27 \text{ mol } O_2 \text{ m}^{-3}$ (Figure 2.27).

What concentration of organic carbon would be required to use up all of that oxygen in respiration (reverse of reaction 2.3)?

From the relevant part of the Redfield ratio ($C : O_2 \approx 106 : 138$), the amount of organic carbon required to use up this oxygen must be:

$$0.27 \times \frac{106}{138} = 0.21 \text{ mol C m}^{-3}$$

Application of two-box model calculations suggests that the average annual Falling Particle flux of organic carbon amounts to about 0.2 mol C m^{-3}. That is close to the estimate of $0.21 \text{ mol C m}^{-3}$ which we have just worked out. Given that particles accumulate at the **pycnocline**, it is not surprising that concentrations of dissolved oxygen can approach zero in this region (Figure 2.29) — such conditions are sometimes described as *sub-oxic*.

Because gases are more soluble in cold than warm water, the deep-water masses formed at the surface in high latitudes are the richest in dissolved oxygen. As the water masses sink and move away from their source regions, the oxygen is progressively used up by marine organisms (including bacteria) in respiration.

QUESTION 2.15 With reference to Figure 2.23, where would you expect to find lowest concentrations of dissolved oxygen in oceanic bottom waters?

Nonetheless, even in the oldest bottom waters, oxygen concentrations rarely fall below about 3 ml l^{-1}, so the deep sea-bed is kept well-oxygenated throughout the major ocean basins. The waters at mid-depths can become very oxygen-deficient (Figure 2.29), and such water layers can be extensive in some oceanic areas. However, it is rare for any part of the water column in the open oceans to become completely devoid of oxygen, but this can — and does — happen in the shallower waters of some continental shelf areas.

Anoxic environments in seawater
Water depths in most continental shelf areas are less than 200 m, so under normal conditions the waters are well-oxygenated because of mixing down of air from the surface by waves and tidal currents. However, in regions of

high productivity, where the amount of sinking organic matter is greater than the available oxygen supply can cope with, the lower part of the water column may become completely depleted in oxygen. Such conditions are described as **anoxic**.

In isolated basins such as the Black Sea, and some Norwegian fjords and Scottish sea lochs, density stratification of the water column and topographic barriers to mixing (sills at the mouths of basins) limit the supply of oxygen-rich bottom waters. The lower part of the water column thus becomes anoxic for at least part of each year. Anoxic conditions can also occur locally in coastal waters, where human activities increase supplies of nutrients and organic matter: examples include fish farms and pulp mills, as well as agriculture, which produces run-off of fertilizers from farm land.

Where water is anoxic, bacteria must use other oxidizing agents to decompose organic matter. Sulphate (SO_4^{2-}) is a major dissolved constituent in seawater (Table 2.1) and when oxygen has been used up, one of the more favourable reactions (in terms of chemical energy), for the decomposition of organic matter is:

$$2CH_2O(s) + SO_4^{2-}(aq) + 2H^+(aq) = 2CO_2(g) + H_2S(g) + 2H_2O \qquad (2.5)$$

As we saw earlier (Section 2.2.3), the **redox** state of the water column can influence the solubility of trace elements that occur in more than one oxidation state. Such elements are sometimes said to be *redox-sensitive*. Manganese is an example of a redox-sensitive element which is more soluble under reducing than under oxidizing conditions (Figure 2.30).

Pore waters within marine sediments can also become sub-oxic and even anoxic at some depth below the sea-bed. This is explored in Chapter 5.

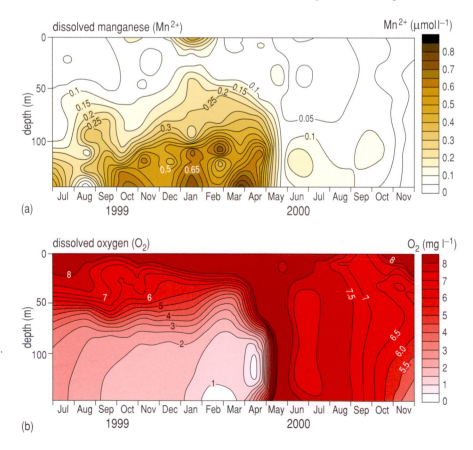

Figure 2.30   Time series showing concentrations of (a) dissolved manganese (Mn^{2+}) and (b) dissolved oxygen (O_2) at a single station close to the centre of Loch Etive, Scotland. The water column was well-stratified below 50 m depth until May 2000, and there are marked changes in O_2 and Mn^{2+} with depth (shown by horizontal contours); note that the deeper waters have lowest O_2 and highest Mn^{2+}. In May 2000, denser oxygenated water flowed in from shallower regions of the loch closer to the sea, leading to mixing of the water column (shown by vertical contours) and lower Mn^{2+}. The contour interval for Mn^{2+} is 0.05 µmol l^{-1}; for O_2 it is 0.5 mg l^{-1}.

2.6 SUMMARY OF CHAPTER 2

1   Most naturally occurring elements have been detected in solution in seawater. Variations in salinity do not affect the overall constancy of composition of seawater with respect to major constituents, most of which behave conservatively. Most minor and trace constituents (and a few major ones, notably carbon, and calcium to a small extent) are non-conservative, because their concentrations are affected by biological processes in the oceans.

2   The oceans are in a steady state. There is an overall balance between the rate of supply of dissolved constituents (including excess volatiles and cyclic salts) and their rate of removal from solution. Residence times range from several tens of millions of years to a few hundred years or less, but most residence times are long compared with the oceanic stirring time. While in the oceans, dissolved constituents participate repeatedly in (mainly) biological cycles before being ultimately removed from solution in seawater into sediments and rocks at the sea-bed. Eventually, these sediments and rocks are accreted to continental margins or returned to the Earth's mantle by subduction at ocean trenches.

3   Much organic matter produced by primary production in surface waters is recycled there, but a proportion sinks out of the photic zone towards the sea-bed. Particulate organic matter ranges from micrometres to centimetres in size. Larger particles are mainly marine snow and faecal material, while the smallest particles are mostly bacteria. In surface waters, the average molar ratio of carbon to the two principal nutrient elements in organic matter — nitrogen and phosphorus — is close to $106:16:1$ ($C:N:P$); this is the basic Redfield ratio. The composition of sinking particles changes with depth: skeletal material (carbonate and silica) dissolves only slowly, but soft tissue is consumed and decomposed by animals and bacteria, and the residue becomes more refractory with depth as nutrients are extracted — the proportion of C relative to N and P in particulate matter increases. Information as to the composition and distribution of particulate matter has come largely from sediment traps.

4   Particulate organic carbon contributes only a small proportion of the organic carbon in seawater ($c.\ 0.05–0.1\ \mathrm{mg\,C\,l^{-1}}$ on average). Most organic carbon is in the form of dissolved organic compounds ($c.\ 0.5–1\ \mathrm{mg\,C\,l^{-1}}$ on average), but these are largely refractory and thus not available for microbial consumption.

5   Formation and decomposition of particulate organic matter is a major regulator of seawater composition, for minor and trace constituents in particular. Elements classified as recycled have concentration profiles resembling those for the major nutrients (nitrate, phosphate, silica). They are taken up by organisms during growth and released back into solution as the organisms are consumed and decomposed on sinking into deeper waters after death. Recycled elements can also be subdivided into biolimiting (concentrations near-zero in surface waters), and bio-intermediate (concentrations only somewhat reduced in surface waters). Which of these categories an element falls into depends greatly on the amounts used in biological production relative to their total concentrations in seawater. Scavenged elements have profiles showing depletion at depth, the result of adsorption onto surfaces of particles (mainly bacteria) and scavenging from the water column by larger sinking particles. The solubility of elements

with more than one oxidation state can vary according to whether conditions are oxidizing or reducing.

6   The two-box model enables first-approximation estimates to be made of the relative amounts of dissolved constituents that are removed into sediments and recycled in the water column. *It can only be applied to elements in the recycled category.* Only a minute fraction of the particulate material containing the biolimited constituents reaches sediments on the sea-bed; the remainder is recycled, mainly above the permanent thermocline.

7   In the deep oceans, there are marked lateral variations in concentrations of biologically active constituents. The pattern of thermohaline circulation results in an overall enrichment of nutrients in the deep waters of the North Pacific relative to the North Atlantic.

8   Transfer of gases, liquids and solids can take place across the air–sea interface. Gas exchange occurs by molecular diffusion and, as wind speed increases, vertical turbulence. Gas exchange is continuous, but at equilibrium there is no *net* flux in either direction. There is approximate equilibrium between atmosphere and ocean for the major gases. Minor gases produced by organisms in surface waters have a net flux from sea to air. Both gases and solids are also transferred across the air–sea interface in precipitation. In this case, the air–sea flux depends on the nature and intensity of precipitation and on the washout ratio; and the sea–air flux depends on the extent to which aerosols are produced.

9   The deep oceans are well supplied with oxygen by deep water masses formed at high latitudes. Dissolved oxygen concentrations decrease as water masses 'age' as they move away from their source regions and marine organisms use the oxygen in respiration/decomposition. The excess of abstraction over replenishment of oxygen reaches a maximum at about 1 km depth, and sub-oxic conditions can develop. High levels of biological production can cause some coastal waters to become sub-oxic or anoxic. In the absence of oxygen, bacteria use oxidizing agents such as sulphate to decompose organic matter.

Now try the following questions to consolidate your understanding of this Chapter.

QUESTION 2.16   The molar ratio in particulate matter dominated by calcareous organisms is $131 : 25 : 16 : 1$ ($C : Ca : N : P$). Why is there more carbon in this ratio than in the 'average' Redfield ratio?

QUESTION 2.17   Samples of surface and deep water from two stations (I and II), one in the Atlantic and the other in the Pacific, were analysed for dissolved zinc (Zn^{2+}) and dissolved cerium (Ce^{3+}). Unfortunately, the samples were incorrectly labelled and it was necessary to inspect the data to determine which came from which ocean. With the help of Figure 2.17, use the data in Table 2.3 to work out which station was in which ocean.

Table 2.3   For use with Question 2.17.

| | Zn^{2+} (10^{-9} mol kg^{-1}) | | Ce^{3+} (10^{-12} mol kg^{-1}) | |
	Station I	Station II	Station I	Station II
Surface	0.8	0.8	66	19
Deep	1.6	8.2	11	6

QUESTION 2.18   How meaningful are the averages in Table 2.1, as far as scavenged elements are concerned?

QUESTION 2.19   To what extent can the two-box model be applied to cyclic salts?

QUESTION 2.20   In making flux calculations for the transfer of DMS from sea to air, the value of ΔC in Equation 2.1 is taken simply to be the measured concentration in surface seawater (see end of Section 2.5.1). Why is that?

QUESTION 2.21   Which of the following statements are true, and which are false?

(a)   All major constituents are conservative in seawater.

(b)   The constancy of composition of seawater extends to recycled constituents.

(c)   Considering the oceans as a whole, a greater volume of water is involved in upwelling than in downwelling.

(d)   Continental shelves are kept well-oxygenated by deep water that has sunk at high latitudes and moved towards the Equator.

(e)   The ratio of sulphate to total salinity (i.e. $SO_4^{2-} : S$) will be less in anoxic than in oxygenated seawater.

CHAPTER 3 — THE ACCUMULATION OF DEEP-SEA SEDIMENTS

Deep-sea sediments of terrigenous and biological origin occur in quite well-defined areas (Figure 1.12). The main factors controlling this distribution include the productivity of surface waters, the water depth, and the supply of terrigenous sediment, large volumes of which will dilute any biogenic components. The geographical separation of calcareous and siliceous sediments is related more to the different solubilities of calcium carbonate and silica and the chemistry of the water column than to the distribution of organisms at the surface.

3.1 BIOGENIC SEDIMENTS

Deep-sea biogenic sediments consist mainly of the skeletal remains of very small planktonic organisms, some only a few μm in size (Section 1.1.1). Theoretically, these remains would take decades or even centuries to reach the sea-floor.

So, how is it that sediments formed of this skeletal debris mainly occur directly below areas of high productivity in surface waters? Why are they not spread all over the sea-bed by ocean currents?

As you read in Chapter 2, the short answer is biopackaging, in the form of marine snow and faecal pellets — it has been estimated that a single faecal pellet can contain as many as 10^5 coccoliths, so biological aggregation is obviously an important way of speedily transferring planktonic debris (along with clays mainly of aeolian origin) to the sea-bed.

3.1.1 THE PRESERVATION OF DEEP-SEA SILICEOUS REMAINS

All seawater is undersaturated with respect to biogenic silica, which means that, in theory, all siliceous remains are subject to dissolution as they sink towards the sea-bed. The solubility of biogenic silica decreases by about 30% for a fall in temperature from 25 to 5 °C, though this decrease is offset slightly in the deep oceans, because high pressure acts to increase solubility slightly. The considerable dissolution and recycling of silica in upper parts of the water column is possible mainly because temperature is higher here, but also because the more undersaturated the water, the greater the rate of dissolution; levels of dissolved silica are depleted in surface waters due to biological production (Figure 2.10(c)). Of the material that is exported from the photic zone however, most reaches the sea-bed intact, so dissolution in the deep sea takes place mainly on the sea-floor. Dissolution of biogenic silica also continues within the sediments, resulting in diffusion of dissolved silica up into bottom waters.

The greater the supply of skeletal debris, the more reaches the ocean floor. Supply is related to productivity, and siliceous sediments on the ocean floor thus underlie areas of high productivity (Figure 1.12). Even in these areas however, only between 1 and 10% of siliceous material escapes dissolution, either in the upper parts of the water column or at the sea-bed, and accumulates to form sediments. This small proportion can be easily swamped, especially where there is a high input of terrigenous sediments, or where the sea-floor is shallow enough for carbonate sediments to be preserved (see Section 3.1.2).

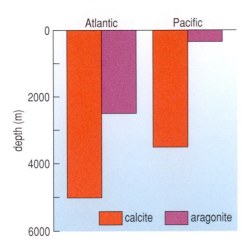

Figure 3.1 Average depth below which $CaCO_3$ is absent in sediments in the Pacific and Atlantic Oceans.

Close to continental margins where inputs of material from the land are high, terrigenous sediments can accumulate at rates as high as a few metres per *thousand* years, enough to obliterate any biogenic component. By contrast, the accumulation rate of a siliceous ooze is of the order of a few metres per *million* years; whereas accumulation rates of pelagic carbonates can reach several tens of metres per million years, which is also quite enough to dilute the siliceous content of sediments beneath regions of high productivity.

As mentioned in Chapter 1, diatom oozes generally predominate in deep waters at high latitudes and in areas of coastal upwelling, whereas radiolarian oozes are found at low latitudes. The belt of siliceous sediments round the Antarctic (Figure 1.12) consists of diatom ooze, whereas that in the equatorial Pacific is of radiolarian ooze. Mixed radiolarian and foraminiferal oozes may occur in the most productive areas.

3.1.2 THE PRESERVATION OF DEEP-SEA CARBONATES

As you saw in Chapter 1, the distribution of carbonate sediments in the ocean basins is far from uniform (Figure 1.12). If it were possible to drain away all of the ocean's water, carbonate oozes would be found draped like snow over the topographic highs of the sea-floor and to be largely absent in the deep basins. The depths below which the two main forms of biogenic calcium carbonate, calcite and aragonite (cf. Section 1.1.1) are generally absent in marine sediments of the Atlantic and Pacific is shown in Figure 3.1. To understand why this should be, we need to examine the inorganic carbon chemistry of seawater.

The carbonate system in seawater

The amount of calcium carbonate that will dissolve in seawater is determined by the equilibrium reaction:

$$CaCO_3(s) \rightleftharpoons Ca^{2+}(aq) + CO_3^{2-}(aq) \tag{3.1}$$

In the oceans, the concentration of Ca^{2+} in solution is virtually constant. As you read in Section 2.1, calcium in seawater departs only very slightly from conservative behaviour; the formation and dissolution of $CaCO_3$ skeletal material has little effect on $[Ca^{2+}]$ (the square brackets, [ ], denote concentration of a dissolved species), so the degree of saturation of $CaCO_3$ is determined by $[CO_3^{2-}]$, which is much more variable.

Figure 3.2 shows the concentration of carbonate ion corresponding to $CaCO_3$ saturation. The saturation curve for aragonite lies to the right of that for calcite as it is less stable and dissolves more readily (Section 1.1.1). The saturation carbonate ion concentration increases with depth because calcium carbonate is more soluble in cold than in warm water and it is also more soluble at high pressure than at low pressure (dissolved calcium and carbonate ions occupy less volume than when combined in solid form). The depth at which calcite and aragonite skeletal material will start to dissolve in the water column can be determined from measurements of seawater $[CO_3^{2-}]$. If the concentration of CO_3^{2-} ions in the seawater lies to the right of the saturation curve, $CaCO_3$ will not dissolve. If the measured value of $[CO_3^{2-}]$ lies to the left of the curve, $CaCO_3$ will dissolve. Figure 3.2 suggests that aragonite should begin to dissolve at a little over 3 km depth, whereas calcite should not start dissolving until nearly 4.5 km depth. The depth at which the measured $[CO_3^{2-}]$ intersects the saturation curve is called the **saturation horizon**.

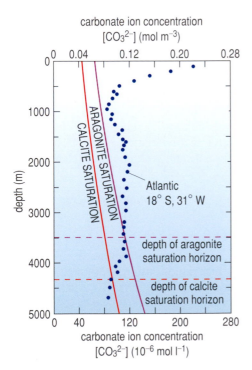

Figure 3.2 Curves of CO_3^{2-} concentration versus depth for saturation of $CaCO_3$ (calcite and aragonite). Dots represent a profile of seawater $[CO_3^{2-}]$ for a station in the Atlantic. Dashed lines denote the depth of the saturation horizon (see text for details).

QUESTION 3.1 Why are the depths of the saturation horizon for calcite and aragonite in Figure 3.2 shallower than the depth below which $CaCO_3$ is absent in sediments in the Atlantic Ocean (Figure 3.1)?

60

An important general point is also illustrated in Figure 3.2: down to a few km depth, seawater is supersaturated with respect to calcium carbonate. Most CaCO$_3$, however, is biogenic: it rarely precipitates spontaneously from seawater due to the inhibiting effects of other ions, such as Mg^{2+}.

While calcium carbonate solubility helps us to explain the depth-dependency of carbonate ooze, and differences between the depth thresholds for dissolution of calcite and aragonite, it cannot explain why carbonates disappear at shallower depths in the Pacific relative to the Atlantic (Figure 3.1), or why the carbonate ion concentration of seawater decreases with depth (Figure 3.2). To do this, we need to take a look at another aspect of the seawater carbonate system.

The carbonate ion is not the only form of inorganic carbon present in seawater. When molecules of carbon dioxide diffuse into surface seawater, some react to produce carbonic acid, H$_2$CO$_3$(aq), but most occur as hydrated CO$_2$ (written as CO$_2$(aq)), where each CO$_2$ molecule is surrounded by water molecules. Because it is difficult to distinguish analytically between CO$_2$(aq) and H$_2$CO$_3$(aq), in practice dissolved carbon dioxide is normally referred to simply as carbonic acid. If we follow this convenient shorthand, we can write the chemical equation for the solution of CO$_2$ gas in seawater as:

$$CO_2(g) + H_2O \rightarrow H_2CO_3(aq) \tag{3.2a}$$

At any particular temperature, the amount of CO$_2$ that diffuses into the water depends on the concentration of CO$_2$ in the atmosphere on the one hand, and the concentration of carbonic acid in the water on the other.

When enough H$_2$CO$_3$ builds up in the water, some of the dissolved carbon is released as CO$_2$ back to the atmosphere in the reverse reaction:

$$H_2CO_3(aq) \rightarrow CO_2(g) + H_2O \tag{3.2b}$$

Eventually, the forward reaction (Equation 3.2a) and the reverse reaction (Equation 3.2b) are occurring at equal rates, and a state of 'dynamic balance' or equilibrium has been established:

$$CO_2(g) + H_2O \rightleftharpoons H_2CO_3(aq) \tag{3.3}$$

The system is, however, more complicated than this. Carbonic acid, like all acids, tends to *dissociate*, i.e. lose a hydrogen ion. A second equilibrium is established between the carbonic acid and its dissociation products, a hydrogen ion (H$^+$) and a hydrogen carbonate (bicarbonate) ion (HCO$_3^-$):

$$H_2CO_3(aq) \rightleftharpoons H^+(aq) + HCO_3^-(aq) \tag{3.4}$$

Furthermore, the hydrogen carbonate ion itself dissociates to form a carbonate ion (CO$_3^{2-}$) and a hydrogen ion. This third equilibrium is written:

$$HCO_3^-(aq) \rightleftharpoons H^+(aq) + CO_3^{2-}(aq) \tag{3.5}$$

So, carbon dissolved in water achieves a state of dynamic chemical equilibrium between the CO$_2$ in the air and that partitioned among three dissolved inorganic carbon compounds (Figure 3.3). Thus, the total amount of dissolved inorganic carbon in seawater (ΣCO_2; Σ = 'sum of') is:

$$\Sigma CO_2 = [H_2CO_3] + [HCO_3^-] + [CO_3^{2-}] \tag{3.6}$$

The relative amount of H$_2$CO$_3$, HCO$_3^-$ and CO$_3^{2-}$ varies with pH; in seawater, most dissolved inorganic carbon is in the form of HCO$_3^-$ ions; the amount of H$_2$CO$_3$ is extremely small (Figure 3.4).

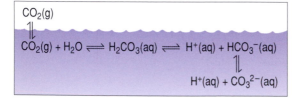

Figure 3.3 Pictorial representation of the carbonate system expressed by Equations 3.3 to 3.5. (Remember that, in reality, H$_2$CO$_3$ is mostly CO$_2$(aq).)

Profiles of ΣCO_2 in seawater take the form shown in Figure 3.5. This is chiefly because carbon dioxide (i.e. H$_2$CO$_3$(aq)) is removed from solution in

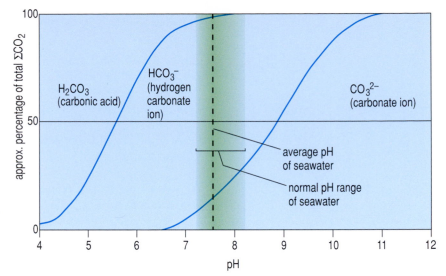

Figure 3.4 Generalized diagram showing approximately how relative proportions of the three principal components in the aqueous carbonate system vary with pH in natural waters. The pH of seawater averages about 7.7 and can range from about 7.2 to 8.2. The positions of the curved lines vary with temperature, salinity and pressure.

surface waters by photosynthesis and then returned to solution in deep water as organic matter is decomposed. If carbon dioxide is added to seawater, pH decreases because carbonic acid is weakly acidic. However, the change in pH is very small because the seawater carbonate system acts like a buffer; it can counter changes in pH by altering the proportions of HCO_3^- and CO_3^{2-} (cf. Figure 3.4). Any increase in dissolved CO_2 ($H_2CO_3(aq)$) can be 'mopped up' by driving the carbonate buffer reaction to the right:

$$H_2CO_3(aq) + CO_3^{2-}(aq) \rightleftharpoons 2HCO_3^-(aq) \tag{3.7}$$

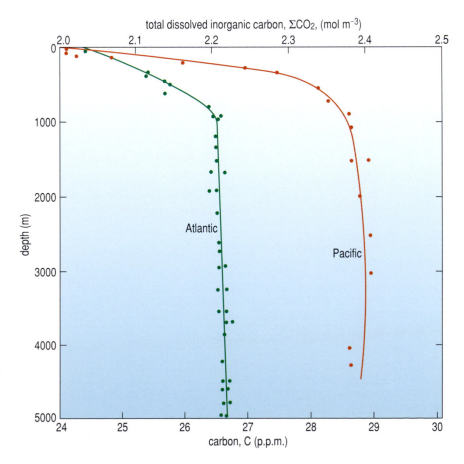

Figure 3.5 Variation with depth of total dissolved inorganic carbon expressed as ΣCO_2 and given in mol m^{-3}, as well as p.p.m. of carbon, in the Atlantic at 36° N, 68° W and the Pacific at 28° N, 122° W. *Note:* to convert mol l^{-1} to mol m^{-3}, simply multiply the concentration by 10^3, e.g. 2×10^{-3} mol l^{-1} becomes 2 mol m^{-3}. Note also that the horizontal axis does not start at zero, and that the concentration increase from surface to deep water is only about 10–20%.

How does this help to explain why $[CO_3^{2-}]$ decreases with depth (Figure 3.2)?

Production of CO_2 by respiration/decomposition lowers the carbonate ion concentration because some CO_3^{2-} is converted into HCO_3^-. The carbonate ion concentration is therefore inversely proportional to ΣCO_2; it is high in surface waters and decreases with depth.

Accordingly, would you expect $CaCO_3$ to dissolve more readily in Atlantic than in Pacific waters, judging from Figure 3.5?

ΣCO_2 is higher in the Pacific than in the Atlantic, so $CaCO_3$ should dissolve more readily in Pacific waters.

The lysocline and the carbonate compensation depth
As most particles are sinking rapidly, they spend only a short time in the undersaturated lower parts of the water column, and there is no chance for significant dissolution to occur. *Dissolution of calcareous material takes place mostly at the sea-bed.*

The depth at which the dissolution of carbonate skeletal material is observed to begin, from analysis of sediment samples from the ocean floor, is called the **lysocline**. Below the lysocline, dissolution occurs at increasing rates, so that there is a progressive decrease in the proportion of carbonate skeletal material preserved in the sediments. The depth at which this proportion falls below 20% of the total sediment is called the **carbonate compensation depth, CCD**. The lysocline and CCD are shallower for aragonite than for calcite, and unless otherwise specified, the terms usually refer to calcite, because skeletal material is much more commonly formed of calcite than aragonite. Figure 3.6 shows the relationship between the calcite content of sediment, the lysocline and the CCD, and also calcite saturation, for a particular area of the eastern equatorial Indian Ocean.

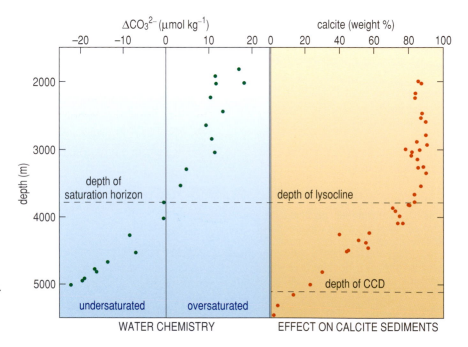

Figure 3.6   Comparison of the degree of calcite saturation (ΔCO_3^{2-}) for the eastern equatorial Indian Ocean with measurements of calcite content (weight %) of depth-distributed modern sediment samples in this region. Degree of calcite saturation is defined as the difference between the actual carbonate ion concentration in the water and the saturation carbonate ion concentration (i.e. $\Delta CO_3^{2-} = [CO_3^{2-}]_{seawater} - [CO_3^{2-}]_{saturation}$). The calcite saturation horizon ($\Delta CO_3^{2-} = 0$) at this location occurs in the water column at a depth of 3800 m. This level coincides with the lysocline as recognized by the beginning of a decline in calcite in the sediments. The CCD in this region is found at a depth of approximately 5000 m; below this depth, the percentage of calcite in the sediments is <20%.

QUESTION 3.2 Figure 3.7 shows profiles of degree of calcite saturation for stations in the Atlantic and Pacific Oceans.
(a) At what depth is the saturation horizon in (i) the Atlantic, and (ii) the Pacific? With reference to Section 2.4, and the text relating to Equation 3.7, briefly account for any difference in the depths that you observe.
(b) How would profiles of degree of aragonite saturation compare with those in Figure 3.7?

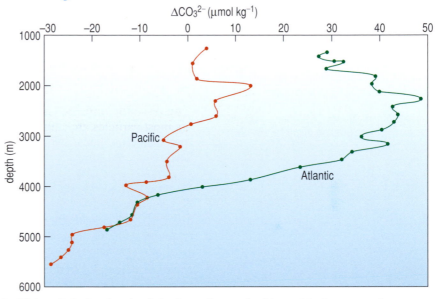

Figure 3.7   Profiles of degree of calcite saturation for stations in the Atlantic and Pacific Oceans. For use with Question 3.2.

In Figure 3.6, the depth of the lysocline coincides with the saturation horizon. In theory, this is always the case, but in practice the lysocline can lie as much as a few hundred metres below the saturation horizon. The reason for this discrepancy is that the $CaCO_3$ content of a sediment sample can be a surprisingly poor indicator of the extent to which $CaCO_3$ dissolution has occurred. For example, the loss (L) of $CaCO_3$ from sediment, expressed as a weight percentage of the total sediment, is given by:

$$L = 100 \times \left(1 - \frac{R_0}{R}\right) \tag{3.8}$$

where R_0 and R are the initial and final values of the non-carbonate (or residual) material. For a sample initially containing 95% carbonate ($R_0 = 5\%$), 50% of the $CaCO_3$ in the sample must be dissolved in order to double the non-carbonate fraction ($R = 10\%$) and reduce the $CaCO_3$ content to 90%. Since the carbonate fraction of particles falling through the deep ocean often approaches 95%, this means that significant loss of $CaCO_3$ can occur before changes in the $CaCO_3$ content of sediments can be detected.

The depth of the CCD is controlled in part by how undersaturated the water is, but also by the nature and flux of debris to the sediments. Coccoliths are more delicate than foraminiferan tests, for example, and the greater the flux, the greater the likelihood of carbonate material being buried before it dissolves. Figure 3.8 shows the depth of the CCD for calcite, as determined from deep-sea sediment samples. The CCD is depressed beneath the Equator on account of the high productivity resulting from equatorial upwelling, and increased flux of calcareous remains to the sea-bed.

It is also apparent from Figure 3.8 that the level of the CCD typically rises towards the continental margins, where biological productivity is in general greater than throughout most of the oceans. This would seem to contradict

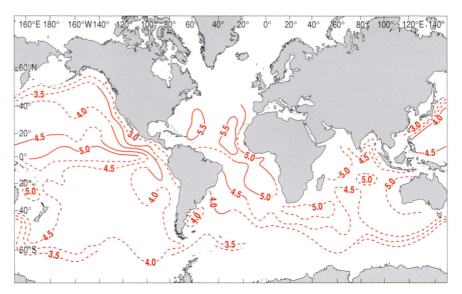

Figure 3.8 Contours (in km) for the calcium carbonate compensation depth (CCD), defined by interpolating boundaries between calcareous sediments and sediments with no or little calcium carbonate. Solid red contours represent more than 20 control samples per 10° square; broken red contours represent fewer than 20 control samples. Note that this map is for the more common calcium carbonate mineral, calcite; the compensation depth for the other variety, aragonite, is very different, as discussed in the text.

the correlation between high productivity and depression of the CCD which we have just established. The reason for the apparent contradiction is that dissolution of calcareous debris is aided by a rich supply of organic matter which is available for consumption by benthic animals and bacteria at the sea-floor, releasing CO_2 into solution at the sediment–water interface. This drives Equation 3.7 to the right, so water becomes more undersaturated in CO_3^{2-} with respect to $CaCO_3$. In these situations (i.e. shallow, highly productive waters), the CCD virtually coincides with the lysocline, whereas in deeper water there is a depth interval of several hundred metres between them (Figure 3.6).

The carbonate system in the oceans is one of considerable complexity, and we have only considered the aspects that most affect the preservation of carbonate sediments here. The basic relationships we have explored can be summarized as follows:

1   The solubility of calcium carbonate in seawater increases with depth (e.g. Figure 3.2).

2   In general, the concentration of ΣCO_2 increases with depth (Figure 3.5). Because the carbonate ion concentration in seawater is inversely proportional to the concentration of ΣCO_2, at some depth the seawater becomes sufficiently undersaturated with respect to calcium carbonate for shelly remains to begin to dissolve, with the result that the $CaCO_3$ content of sea-bed sediments is noticeably affected (e.g. Figure 3.6). This is the lysocline.

3   At some greater depth, very little (<20%) $CaCO_3$ survives in surface sea-bed sediments. This is the carbonate compensation depth, the CCD (e.g. Figure 3.6).

4   Calcite is more stable than aragonite, so the saturation horizon, lysocline and CCD are all shallower for aragonite than for calcite (e.g. Figure 3.2).

3.2   TERRIGENOUS SEDIMENTS

During continental weathering, rocks are fragmented into progressively smaller pieces by physical weathering; their minerals are decomposed and reconstituted into new minerals (chiefly clay minerals) by chemical weathering, which also releases ions into solution. Breakdown by physical weathering is the dominant process in cold, high-latitude regions, whereas chemical weathering is dominant in the warm, wet low latitudes. During the process of transport to the oceans, physical breakdown continues and, if the sediment is water-borne, so also does chemical attack. At the same time, coarse sediment (boulders, gravel, coarse sand) is progressively sorted from fine sediment and deposited in continental areas or in shallow coastal water. In general, except in the polar regions where ice transport is important, only grains finer than gravel (less than 2 mm diameter; sand, clay and silt) reach the outer **continental shelf** and, eventually, the ocean basins.

By the time terrigenous sediment reaches the deep ocean basins, physical and chemical breakdown are usually well advanced. The particles consist mostly of fine quartz grains (which are hard, and resistant to chemical weathering) and clay minerals (the decomposition products of minerals such as feldspar and mica).

QUESTION 3.3   Look back at Figure 1.12. Which sort of land-derived sediment represented on the map is likely to contain coarse material showing the least evidence of chemical breakdown?

Glacial sediments are latitudinally localized in their occurrence and contribute about 10% of the sedimentary materials that reach the oceans. Globally, the largest input of sediments comes from rivers which provide more than 80% of the total. Over two-thirds of the sediment supplied by rivers to the oceans is derived from the great rivers of southern Asia and from the larger islands in the Pacific and Indian Oceans.

By no means all of the sediment brought in by rivers reaches the deep oceans. Round the Pacific margins, although some is deposited on the narrow continental shelves, much is trapped in **marginal basins** behind island arcs, or is deposited in trenches. Round the Atlantic and Indian Ocean margins, sediment is generally deposited on the continental shelf. Material on the continental shelf is reworked by waves and currents; if these are strong enough, then fine material is resuspended and may escape from the shelf edge, to be carried down the **continental slope**.

There are two important mechanisms by which this suspended material is transported.

1   **Lutite flows**   This is the name given to low-density turbidity currents in which concentrations of suspended material do not exceed a few tens of milligrams per litre. (The term lutite simply refers to material made up mainly of clay-sized and silt-sized particles.) The excess density imparted to the water by these low concentrations is not enough to drive a current through the density stratifications of the full open ocean water column and down to the deep sea-floor; deposits from lutite flows are presumably restricted to the upper parts of continental slopes.

2   **Cascading**   Density stratification can develop in mid-latitude shelf seas during the summer months. This is broken down in winter by the combined action of winds and tidal currents, and the whole water column can be cooled and mixed, provided the sea is shallow enough. When this happens near the shelf edge, the shelf water may become colder and denser than the adjacent slope water, and will cascade off the shelf and down the slope, carrying suspended sediment with it. If cascading occurs during a stormy season, it can be an effective mechanism for carrying away resuspended sediment, but it is unlikely to extend far down the continental slope because the deeper layers of the ocean are usually denser than the cascading water.

Do any of the processes we have considered so far suggest a way of transporting large amounts of sediment into the ocean basins?

The short answer is no. Material carried directly over the shelf edge does not travel further than the upper continental slope, either in lutite flows or cascading water. In canyons, both lutite flows and cascading can increase the movement of suspended material that is already moving downwards; but by themselves these processes do not seem capable of moving large amounts of sediment from shelves to ocean basins, even allowing for increased transport during storms.

3.2.1 TURBIDITY CURRENTS AND OTHER GRAVITY FLOWS

Much of the material that accumulates on the upper continental slope is in an unstable situation and likely to move down the slope. Such movements can be classified according to the degree of internal deformation of the mass of sediment that is moved. They range from slides and slumps (in which deformation is minimal) through debris flows (with moderate deformation) to turbidity currents (in which the sediment is dispersed as a turbulent suspension in seawater). Slides and slumps can develop into debris flows and turbidity currents, depending on a variety of factors, such as the volume and nature of the sediments involved, their degree of compaction and water content, the angle of inclination ('dip') of sedimentary layering and the gradient of the continental slope, the intensity of the triggering mechanism, and so on.

Slides, slumps and debris flows

Slides and *slumps* result from mechanical failure along inclined planes allowing large masses of sediment to slip (in a manner very similar to landslips). These masses may be tens to hundred of metres thick and hundreds to thousands of metres long and wide (Figure 3.9). There is

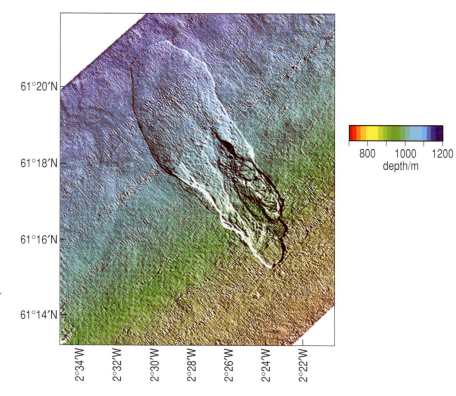

Figure 3.9 Three-dimensional seismic image of the sea-bed in the vicinity of the AFEN submarine slide, which is located approximately 95 km north-west of the Shetland Islands in the Atlantic Ocean. The head of the slide is at a water depth of ~830 m and its base is at ~1120 m. The slide is up to 20 m thick and 4.5 km wide, and its overall length (from its head to its base) is 12 km. Failure of the slope is thought to have been triggered by seismic activity focused on the nearby Victory transfer zone.

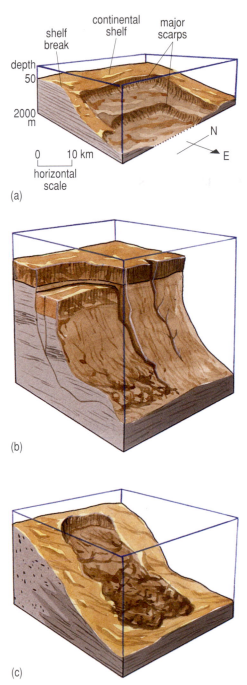

(a)

(b)

(c)

Figure 3.10   (a) General schematic view of scars from the Currituck slides which occurred in the late Pleistocene (~0.1 Myr ago) to the north-east of Cape Hatteras (36° 30′ N, 74° 30′ W). The smaller slide occurred after the larger one, and the total volume of sediments involved was about 130 km³. In this case, failure occurred parallel to the sedimentary layering. In contrast, slumps (b) occur along curved failure surfaces that cut across the sedimentary layers, and the slumped sediments retain some coherence, except at the base of the slump. However, in debris flows (c), all coherence of the original sedimentary layering is lost.

relatively little deformation of the sediment — so that original layering remains detectable when it comes to rest — but the movement gives rise to a bulging hummocky topography at the base of the continental slope and on the upper part of the continental rise. Scars are left higher up on the continental slope where the slides and slumps originated — analogous features can be observed above many landslips (Figure 3.10(a) and (b)). Failure may occur on slopes of as little as 2° and may be triggered by earthquakes, or may simply occur after the relatively rapid accumulation of sediments that are unstable because of their high water content. The distinction between submarine slides and slumps is not always obvious (nor is it usually important) and you may find the terms used interchangeably in the literature.

One of the world's largest known slides is found in the Storegga area off the west coast of Norway (about 62° N). Three slides are recognized; the first occurred between about 30 000 and 50 000 years ago, the other two some 6000 to 8000 years ago. The slides also triggered debris flows and turbidity currents, and as a result a total of nearly 6000 km³ of sediments was transported from the continental shelf break down to water depths as great as 3500 m and over distances up to 800 km. They are believed to have been set off by a combination of earthquake activity and the release of gas hydrates from the sediments (see Section 5.2.2).

Debris flows (Figure 3.10(c)) originate for the same reason as slides and slumps, and they involve the sluggish movement of a mixture of sediments down the slope. Internal deformation is sufficient to obliterate all traces of original layering, but the material retains some coherence. Debris flows typically comprise material that ranges from boulders to clay particles (Figure 3.11), and they can flow for considerable distances over slopes of as little as 0.1°. They are dominated by the cohesive nature of the clay matrix, the strength of which is mainly responsible for supporting the

Figure 3.11   Section through a debris flow exposed in a road cutting in NW Ecuador.

larger particles. Several examples of large debris flows have been discovered along continental margins. Flows recorded off north-west Africa (about 25° N), for example, involved the displacement of about 600 km^3 of sediment; this is as much terrigenous sediment as the nearshore basins off southern California have received from turbidity currents in a million years.

Slides, slumps and debris flows in general contribute to development of the continental rise at the base of the continental slope. Transport of sediment out onto the abyssal plains is mainly by turbidity currents.

Turbidity currents

Turbidity currents are high-velocity density currents, which carry larger amounts of suspended sediment than lutite flows and are therefore denser.

They are considered to be responsible for incising **submarine canyons**, for the breaking of submarine cables, and for the formation of submarine fans and the deposition of sand layers on **abyssal plains**; indeed, abyssal plains are built mainly from successive deposits from turbidity currents which have in places accumulated to thicknesses in excess of 1000 m.

Turbidity currents are probably triggered by the same mechanisms that set off slides, slumps and debris flows, and they are formidable transporters of sediment from the continental slope to the deep oceans. These huge masses of sediment-laden water can travel at up to 90 km hr^{-1} (25 m s^{-1}), carrying up to 300 kg m^{-3} (300 g l^{-1}) of material in suspension, including gravel-sized particles as well as large quantities of mud, and they can transport this material up to 1000 km from the source. Their flow is not necessarily confined to canyons or channels, but may advance over a broad front, only parts of the flow being channelled.

Important indirect evidence about the speeds of turbidity currents has been obtained by studying the breaks in submarine cables crossed by the currents. The most often-quoted example is the turbidity current that resulted from an earthquake on the Grand Banks, Newfoundland, in November 1929 (Figure 3.12). Following the earthquake, 12 submarine telegraph cables were broken in at least 23 places over a period of about 12 hours. At first, it was assumed that the earthquake had caused the damage by itself, and it was not until 1952 that the breaks were related to a turbidity current. It appeared that the earthquake had triggered a massive slump on the continental slope, and this broke submarine cables in the immediate vicinity. A turbidity current deriving from the slump travelled on down the slope, partly along shallow channels but mainly as a broad-fronted flow.

It is believed that this particular flow extended about 800 km from its source, out across the abyssal plain, before it stopped, and that at its fastest it reached between 40 and 55 km hr^{-1} (between 11 and 15 m s^{-1}). There is evidence that many turbidity currents travel even faster, commonly attaining maximum speeds of around 25 m s^{-1} (90 km hr^{-1}), slowing to less than 10 m s^{-1} (36 km hr^{-1}) at the extremities of submarine fans and to only 0.1–0.2 m s^{-1} (1 km hr^{-1}) on the outer part of abyssal plains.

Most deposition from turbidity currents occurs near the base of the continental slope. The comparatively sudden decrease in current velocity associated with the decrease in the gradient of the sea-bed leads to rapid deposition. The sediment is built up into lobe-shaped **submarine fans** that

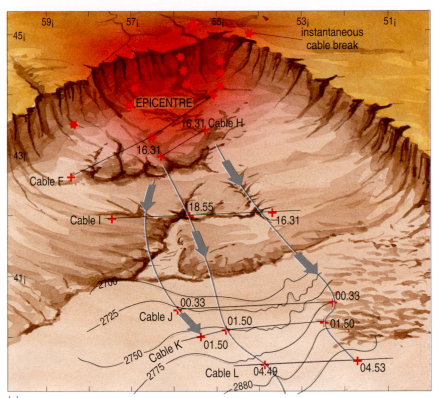

(a)

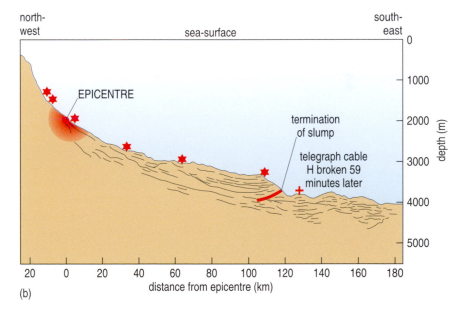

(b)

Figure 3.12 (a) Combined map and perspective view of the area of the Grand Banks earthquake in 1929, showing the epicentre, the cables broken by the turbidity current(s) and the times at which they were broken. The first cables were broken an hour after the earthquake, which occurred at about 15.30 hours. The sea-bed lies at about 2000 m depth in the vicinity of cables F and H. Red stars in the red shaded area show cable breaks due to the earthquake itself (see (b)), and red crosses show breaks due to the turbidity current afterwards.

(b) Interpreted seismic reflection profile of the Grand Banks area, showing the configuration of the initial slump in relation to the earthquake epicentre and the first cable break by the turbidity current. Stars and crosses as in (a).

70

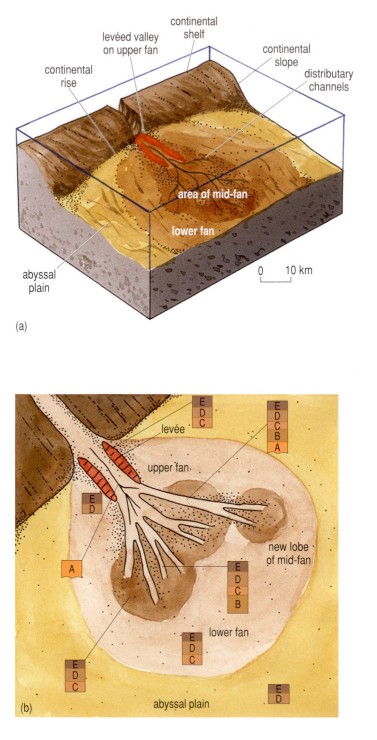

(a)

(b)

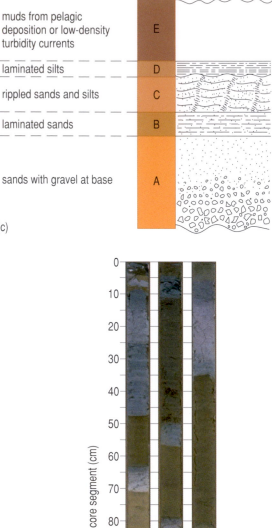

muds from pelagic
deposition or low-density
turbidity currents — E

laminated silts — D

rippled sands and silts — C

laminated sands — B

sands with gravel at base — A

current speed decreasing

(c)

depth in core segment (cm)

(d)

Figure 3.13 (a) General view of a submarine fan. The vertical scale is greatly exaggerated.
(b) Distribution of different types of turbidite sequences on a submarine fan. For key to the letters, see (c).
(c) Schematic representation of the 'ideal' or complete turbidite sequence. The wavy line at the base of the sequence indicates an *erosion surface*, because turbidity currents erode the surfaces over which they travel as well as carving out canyons. The upper wavy line is the base of the overlying sequence.
(d) Photograph of a segmented sediment core from the Madeira abyssal plain. The top segment of the core (recovered from 309.9 m below sea-floor) is at the upper left, the bottom segment (total length of core *c.* 4.5 m) is at the bottom right. The sediment sequence comprises an alternation of layers of rapidly deposited organic-rich olive-green turbidites, and much more slowly deposited pale grey to brown pelagic ooze and clay.

may be tens to hundreds of kilometres across (Figure 3.13(a)), and which contribute to development of the continental rise, along with slides, slumps and debris flows. Subsequent turbidity currents flow out over the surface of the fan, carving channels in its surface and extending the deposits seawards. The coarsest sediment — gravel and coarse sand — is found in the channels on those parts of the fans closest to the foot of the continental slope (Figure 3.13(b)). However, at the outer part of the fan, where current velocity is greatly diminished, finer sands and silts are deposited. The finest clays remain in suspension until eventually they settle out in the intervals between turbidity currents (Figure 3.13(c) and (d)). It is probable that some of this fine sediment is carried out into the deepest parts of the ocean basins to be deposited along with pelagic sediments.

Thus, the typical products of most turbidity currents are the successions of gravels, sands, silts and muds (clays) known as **turbidite sequences** (Figure 3.13(c)). However, it is clear from Figure 3.13(c) that different parts of a sequence will be found in different parts of the submarine fan — it is rare for the full sequence to be recovered in a single core. It is also clear from Figure 3.13(b), that, in their *terminal stages* (i.e. out on the abyssal plain), turbidity currents are typically dilute suspensions containing sediment concentrations of 1–2 kg m^{-3}. This is still a good deal more concentrated than lutite flows, but nonetheless represents a great reduction in sediment load compared to the early stages of a turbidity current flow.

Increases in rates of accumulation of sediments on abyssal plains during the past 15 Ma or so, especially in the North Atlantic and North Pacific, have been linked to an increase in turbidite activity. Sea-level fell with the onset of glaciations in the Northern Hemisphere, so rivers flowing across broad coastal plains (formerly submerged continental shelves) dropped much of their sediment load close to what is now the shelf edge. Influxes of fresh sediment would have been sufficient to trigger slumping and to generate turbidity currents. Thus, it seems that turbidite activity can be used as an indicator of periods of climate change. Deep-sea sediments contain a wealth of information about past climates; this is the subject of the next Chapter.

3.3 SUMMARY OF CHAPTER 3

1   Biogenic sediments are most abundant where productivity in surface waters is high and terrigenous sediments are scarce. The geographical separation of carbonate and siliceous sediments is related to the preservation potential of the planktonic organisms and the chemistry of the water column. Biological aggregation is important in the transport of planktonic skeletal remains to the sea-bed.

2   Seawater at all depths is undersaturated with respect to silica, and the preservation of siliceous sediments depends on the survival of siliceous skeletal debris as it descends the water column. The likelihood of preservation is greater once the remains have survived descent through surface waters. The chances of siliceous sediments accumulating are greatest where surface productivity is high, and where water depths are great, so that dilution by terrigenous or calcareous material is low. Siliceous sediments are most abundant at high latitudes in the Pacific Ocean, in equatorial regions of both the Pacific and Indian Oceans, and in coastal upwelling areas.

3   Carbon dioxide gas dissolves in seawater to form dissolved CO_2 (commonly referred to as carbonic acid, H_2CO_3) and hydrogen carbonate

(bicarbonate) and carbonate ions. Hydrogen carbonate and carbonate ions are also supplied to the oceans by rivers. Total dissolved inorganic carbon (ΣCO_2) increases in deep water chiefly because of the decomposition of organic matter in the water column (liberating CO_2, which goes into solution to form carbonic acid). The greater the concentration of ΣCO_2, the more carbonic acid in the water so the more likely calcium carbonate is to dissolve.

4   The solubility of calcium carbonate increases with depth in the oceans. Surface waters are supersaturated with respect to calcium carbonate, and deep waters are undersaturated.

5   The depth at which dissolution of calcium carbonate in sediments is observed to commence is called the lysocline. It lies at or below the depth where the water is shown from measurements to be saturated with respect to calcium carbonate (the saturation horizon). The carbonate compensation depth (CCD) is defined as the depth at which the carbonate content of the sediments is 20% or less. It tends to be depressed (deeper) beneath areas of high biological productivity in the open oceans. Both the lysocline and CCD are shallower for aragonite than for calcite.

6   Terrigenous sediments are the products of physical and chemical weathering at the land surface. They are transported to the continental shelves and redistributed by waves and currents. Fine sediment is resuspended at the shelf edge by waves and currents. It escapes down the continental slope in low-density suspension (lutite flows) and by cascading. Where the sediments at the shelf edge are unstable, slides, slumps or debris flows occur (along with turbidity currents), carrying large quantities of coarse and fine sediment down to the continental rise.

7   Turbidity currents are the most important means whereby terrigenous sediment is transported from the continental slope out to the abyssal plains. They can travel at speeds of several tens of kilometres per hour.

8   The passage of most turbidity currents is marked by characteristic deposits known as turbidite sequences which build up submarine fans at the base of the continental slope. The terminal stages of turbidity currents are commonly dilute suspensions of silt and mud.

Now try the following questions to consolidate your understanding of this Chapter.

QUESTION 3.4   Explain briefly why neither calcite nor aragonite is likely to be accumulating to any significant extent on the sea-bed at the station represented by the profile in Figure 3.2.

QUESTION 3.5   Give two reasons why CO_2 will be released to the atmosphere where deeper water upwells to the surface in tropical latitudes.

QUESTION 3.6   In Figure 3.2, there is a distinct minimum in the $[CO_3^{2-}]$ profile at about 1000 m depth, superimposed on the overall downward decrease. Can you offer an explanation for this minimum?

QUESTION 3.7   (a) Would you expect submarine fans to be better developed along the margins of the Atlantic or the Pacific Ocean?

(b) The rise of the Himalayas is believed to have accelerated in the past 3 Ma. Would you expect the number of turbidites building up on the Indus and Bengal fans to have increased or decreased over this period?

QUESTION 3.8   Most sediment deposition on the abyssal plains comes from normal pelagic sedimentation (i.e. deposition from the water column). True or false?

CHAPTER 4

DEEP-SEA SEDIMENTS AND PALAEOCEANOGRAPHY

As recently as the late 1960s, the deep sea was regarded by some as an unchanging primordial environment and the most constant feature on the surface of the Earth throughout most of geological time. While the HMS *Challenger* expedition (Section 1.1) provided information about the present-day distribution of deep-sea sediments, an historical perspective was added only in the mid-1940s following the invention of the **piston core**. This was first deployed on the Swedish Deep-Sea Expedition (1947–1948), and analysis of the recovered sediment cores revealed that over time there had been significant variations in the accumulation of carbonate material, which could be correlated with glacial and interglacial conditions. Although sedimentary studies such as this were important in demonstrating that the oceanic environment had in fact changed with time, the science of **palaeoceanography** (the study of ancient oceans) was really founded in the 1950s, when Cesare Emiliani (Figure 4.1) published a series of three classic papers that outlined past variations in sea-surface temperature as derived from the analysis of oxygen isotopes in the $CaCO_3$ shells of foraminiferans (see Section 4.3.1). These papers demolished the notion that the marine environment is constant over periods of thousands of years (Figure 4.2).

Piston cores can recover material up to a few hundred thousand years in age. Much longer cores could be recovered only after the start of the Deep Sea Drilling Project (DSDP) in 1968. DSDP ended in 1983, and continued as a multinational effort with the Ocean Drilling Program (ODP, 1985–2003), which has now been superseded by the Integrated Ocean Drilling Program (IODP, 2003–). IODP utilizes two drillships, along with additional drilling platforms as required

Figure 4.1 Cesare Emiliani (1922–1995). Emiliani was born in Italy; this photograph was taken in the early 1950s when he was working at the University of Chicago. (*Photo: Robert Ginsburg.*)

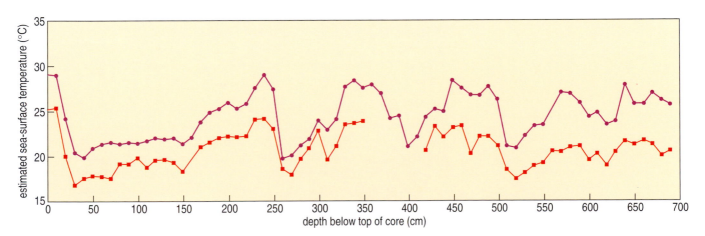

Figure 4.2 Emiliani's first measurements of the variation in sea-surface temperature with core depth, as determined from oxygen-isotope analysis of planktonic foraminiferans, for a core obtained from the Caribbean. (i) *Globerigerinoides sacculifera* (purple circles); (ii) *Globigerina dubia* (red squares). Note that the value determined for sea-surface temperature varies with species. Radiocarbon dating of this core gave an age of 27 600 ± 1000 years for sediment at a depth of 73.5 cm below the top of the core.

(Figure 4.3). Core material recovered by DSDP, ODP and IODP has confirmed that there have been changes in sea-surface temperature in the past, and it has shown that sediments accumulating on the sea-floor in fact record a plethora of other conditions (and events) in the overlying waters, such as biological productivity and patterns of ocean circulation. This is explored in the following Sections: first though, we will investigate how the sediment record has helped to unravel the history of the evolution of the ocean basins.

Figure 4.3   The IODP drillship *Chikyu*, launched by the Japanese in 2004. '*Chikyu*' is Japanese for 'Earth'.

4.1   EVOLUTION OF THE OCEAN BASINS

The sea-floor sediment record extends back to ~180 million years ago at the most.

Why is that, when the oldest rocks on Earth are around 3850 million years old?

The reason is that the ocean basins are relatively short-lived features of this planet. An individual basin grows from an initial rift, reaches a maximum size, then develops subduction zones around the margins, begins shrinking and ultimately closes completely. This cycle takes, on average, about 200 million years (Ma): no oceanic crust older than about 180 Ma is known from the present oceans. As sediments accumulate on top of oceanic crust, they too can be no older than about 180 Ma.

The thickness of sediments increases with distance from spreading axes, and with increasing depth either side of an axis.

What is the reason for this?

The further crust is from the ridge, the older it is and the more time has elapsed for sediments to accumulate; furthermore, depth increases with age according to the **age–depth relationship** (Figure 4.4). Near spreading axes, sediments are no more than a metre or two thick, even in depressions in the rugged topography, except where an axis lies close to land (e.g., in the Gulf of California). By contrast, in abyssal plain areas, sediment thicknesses of a kilometre or so are commonplace. The continental shelf–slope–rise region may be blanketed by more than 10 km of sediment.

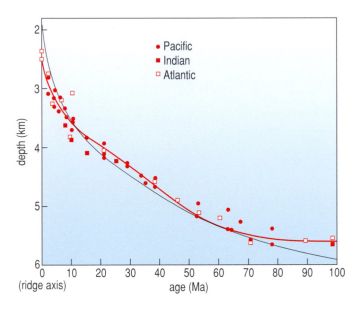

Figure 4.4   Observed and theoretical relationship between the depth to the top of the oceanic crust and its age. The red line is a best-fit curve through observed points. The fine black line is a theoretical elevation curve, calculated on the assumption that the increase of depth with age is due to thermal contraction of the lithosphere as the plate cools on moving away from the ridge axis.

Distance from a spreading ridge may also affect the *type* of sediment accumulating.

QUESTION 4.1   Examine Figure 4.5 and, recalling what you read in Chapter 3, explain why: (a) the main sedimentary component at the sea-floor at locality 1 is calcium carbonate; (b) the carbonate ooze at locality 2 is overlain by pelagic clay; (c) the main sedimentary component *at the sea-floor* at locality 3 is siliceous ooze.

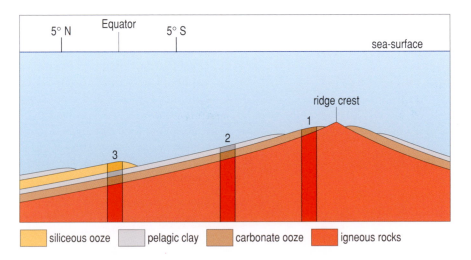

Figure 4.5   A simplified cross-section (perpendicular to the ridge crest) of the sediments in part of the Pacific Ocean. The main components of the sediment types are given in the key. (For use with Question 4.1.)

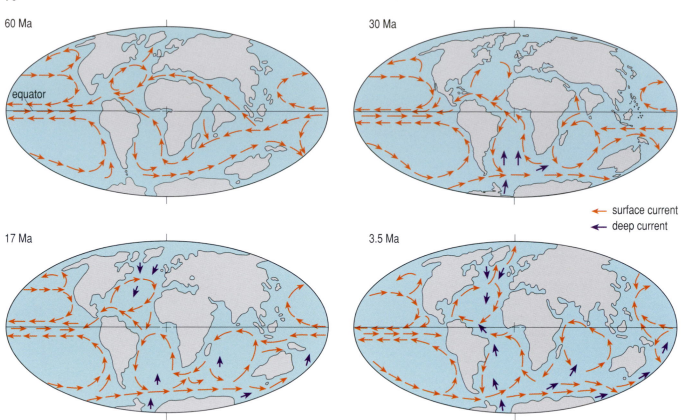

surface current
deep current

Figure 4.6   Changing continental configurations since 60 Ma, and postulated ocean currents (surface: thin orange arrows; deep: thick blue arrows). The maps show two key events: (i) the opening of an oceanic gateway between South America and Antarctica 25 to 30 Ma ago, allowing the Antarctic Circumpolar Current to develop; and (ii) the closing of the gateway between Central and South America at around 3.5 Ma, isolating the Atlantic and shutting down equatorial circulation between it and the Pacific.

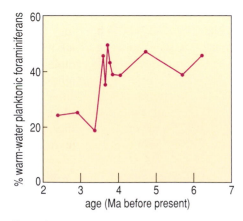

Figure 4.7   Changes in the percentage of warm-water planktonic foraminiferans (relative to total foraminiferal assemblage) over the past ~7 Ma, in sediments recovered from off the coast of NW Ecuador in the Pacific Ocean.

The sediment record can additionally provide information as to the changing configuration of the continents, which in turn affects wind and current patterns (Figure 4.6). By way of example, we will examine the effects of the closure of the gateway between Central and South America at around 3.5 Ma ago. The closure severed communication between the equatorial Pacific and the Atlantic Ocean and Caribbean Sea, and may have played a role in the initiation of the **glaciation** of the Northern Hemisphere.

Figure 4.7 shows that off Ecuador the relative amount of planktonic foraminiferans that prefer to live in warm water decreased sharply in sediments laid down around 3.95 to 3.35 Ma ago, indicating that surface waters in the equatorial Pacific cooled after the closure of the gateway. The nature of the sediments has also changed. Before ~3.5 Ma, they were dominated by silts and sands, but since 3.5 Ma the sediments consist mainly of diatom ooze. As you read in Chapter 1, diatom oozes generally accumulate today in areas of coastal upwelling; upwelling brings cooler, deeper waters to the surface, so the sediments, and the foraminiferan tests preserved within them, are telling us that closure of the gateway between Central and Southern America initiated upwelling off the coast of Ecuador.

Deducing the evolution of the ocean basins is just one aspect of palaeoceanography. The study of palaeoceanography has expanded rapidly in recent years; the main reason for this is that it has become clear that changes in characteristics of the ocean (such as temperature distribution, circulation and productivity) are linked to changes in global climate. A good example of this is shown in Figure 4.8: the variation in the oxygen-isotope composition of surface seawater over the past 400 000 years tracks the variation in atmospheric carbon dioxide and temperature recorded by

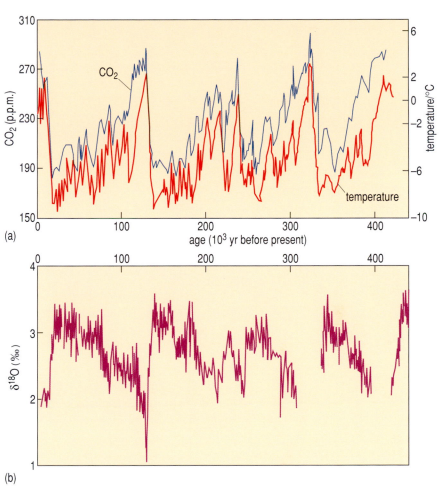

(a)

(b)

Figure 4.8 (a) Record of the last 400 000 years of atmospheric CO_2 and temperature variation from the Vostok ice core, Antarctica. Also shown is (b), the corresponding record of planktonic foraminiferal $\delta^{18}O$ (see Section 4.3.1) from sediments recovered from the South Atlantic (40° 56′ S, 9° 54′ E). Note the close correlation between these records; this tells us that there is some kind of link between atmospheric CO_2, temperature and surface seawater $\delta^{18}O$. The exact nature of this relationship is the subject of ongoing scientific debate.

ice cores from the Antarctic. If we can understand these links between ocean and climate, then we will be better able to predict the effects of human activity on future climate.

4.2 USE OF PROXIES IN PALAEOCEANOGRAPHY

How can we obtain information about the ancient oceans when, for example, scientific observations of sea-surface temperature go back no more than a couple of hundred years, and no laboratory holds samples of glacial seawater? Instead of direct measurements, palaeoceanographers must rely on *proxies* (substitutes) — things that can be measured in sediments (or ice) laid down in the past, and which have responded systematically to changes in important (but now unmeasurable) variables, like salinity. One of the most useful sediment proxies is the **microfossil assemblage**: marine organisms, plant and animal, have their particular ecological preferences, which are related to the environmental conditions in the water in which they lived, such as temperature, salinity and nutrient abundance. Another useful proxy is the composition of microfossil shells, organic matter and other sediment components, in terms of the chemical elements and the relative proportions of the different isotopes of those elements.

The ideal sediment proxy should correlate with a single variable, and this correlation should be preserved in perpetuity in the sedimentary record.

Generally, this isn't the case; the correlation is often imperfect and several more or less independent variables will determine the behaviour of the proxy, while chemical changes within the sediment can overprint the primary correlation between the variable and the proxy. The best way to overcome these problems is to estimate a variable using more than one proxy: the 'multiproxy approach'.

The relationship between a proxy and a variable is calibrated by comparing proxy data for surface (modern-day) sediments with corresponding data collected in the overlying water column.

What critical assumption would you be making here?

You would be assuming that the calibration would also have applied in the past; in, say, glacial times, for example.

QUESTION 4.2   Explain how (a) emplacement of a turbidite, and (b) a low sedimentation rate, can affect the quality of a palaeoceanographic record.

In the following Section, you will learn about some proxies, and what they can tell us about past oceanic conditions and long-term climate change. It will become clear that palaeoceanography is an emerging science, and that much more research is needed into the development and calibration of proxies.

4.3   RECONSTRUCTING PAST OCEANS: INSIGHTS AS TO THE CAUSES OF CLIMATE CHANGE

Figure 4.9 represents a qualitative estimate of the changing mean global temperature for the past ~200 Ma (the maximum age of deep-sea sediments), in relation to that of the present day (marked by a dashed line), which is about 15 °C. Throughout most of this time, mean global temperature has been consistently higher than that of today, and the Earth is informally referred to as having been in a 'greenhouse' condition.

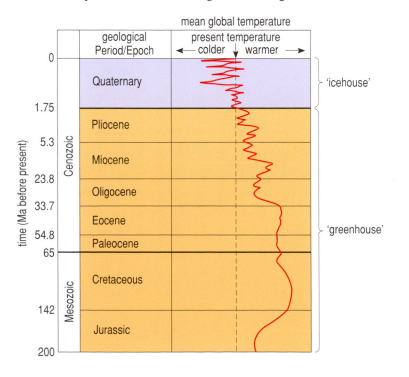

Figure 4.9   Schematic representation of changes in mean global temperature relative to that of the present day (about 15 °C), for the past 200 Ma, based on an array of geological evidence. Note that the time axis is *not* linear, but has been adjusted to show more detail for recent geological periods.

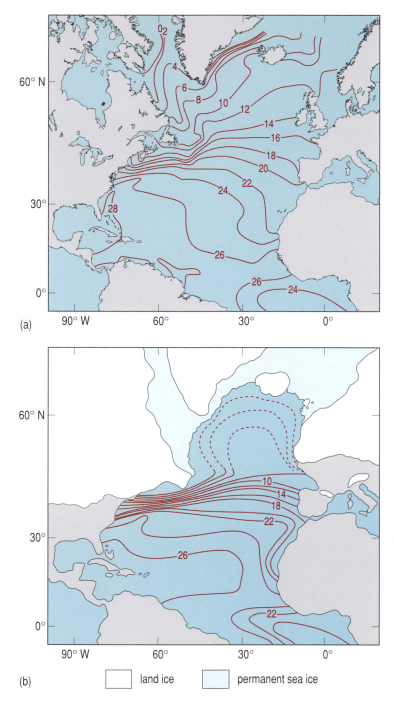

The Quaternary, the present geological era, is however characterized by 'icehouse' conditions, with huge permanent ice-sheets present over Antarctica and Greenland. Another feature of the Quaternary is the dramatic and apparently cyclic changes in the Earth's climatic regime. These cycles involve the growth and retreat of yet further large ice-sheets over much of North America and northern Eurasia (including the British Isles) and elsewhere, and they are referred to as **glacial/interglacial cycles**.

Today, the Earth is in an interglacial period. If we travelled back in time 18 000 years, to the height of the last glacial period, we would hardly recognize our planet; environmental conditions were very different (Figure 4.10).

Many parts of the Northern Hemisphere that are densely populated today were frozen wastes: glaciers reached the sites of New York and Berlin, and the ground where London and Paris would become established was permanently frozen. Ice sheets over 3 km thick covered much of North America and Scandinavia. Winds blowing down from the ice sheets reached speeds of over 100 km per hour. Much of western Europe was largely treeless with a tundra-like landscape, and much of the northern part of the North Atlantic Ocean was frozen, although the Norwegian Sea remained seasonally ice free. Sea ice extended down to the latitude of the British Isles, and even during the summer months, floating ice and large icebergs would have been common between Ireland and Newfoundland.

The isotherms (lines of equal temperature) in Figure 4.10 show that the distribution of sea-surface temperature in the northern Atlantic 18 000 years ago was markedly different from what it is today. At around 40° N, the isotherms were much more crowded together during glacial times as a result of the southward expansion of the high-latitude cold belt. Note, however, that the zone of warm water in equatorial and tropical latitudes (up to about 30° N) was apparently not much narrower at the height of the glaciation than it is now. Off north-west Africa, the relatively low temperatures were even more marked 18 000 years ago than they are today, suggesting that upwelling there was even more powerful in glacial times than at present.

Figure 4.10 Maps comparing the distribution of ice and sea-surface temperatures in the northern Atlantic summer (a) at the present time, and (b) 18 000 years ago. Isotherms are in °C.

QUESTION 4.3 What do the isotherms in Figure 4.10 suggest about the flow of the Gulf Stream 18 000 years ago?

We now move on to look into how maps of past oceanic conditions, such as that shown in Figure 4.10(b), are determined. For the most part, we will use examples from the past ~2 Ma, because the climatic, environmental and ecological processes that have operated during this time, though subject to rapid change, are similar, but not always identical, to those of the present. If we can understand the mechanisms that have driven climate change over the past ~2 Ma, then we may be able to predict future climatic and ecological changes.

4.3.1 SEA-SURFACE TEMPERATURE

Obtaining accurate and detailed records of past sea-surface temperature (SST) is essential for understanding how Earth's climate has operated in the past, because SST drives atmospheric circulation, generating winds and weather, and it influences evaporation, regulating the water cycle and precipitation patterns. Furthermore, it affects seawater density which, in turn, determines patterns of deep-water circulation.

The most widely used proxy for SST is that which was mentioned at the start of this Chapter: the oxygen-isotope composition of the calcium carbonate tests of planktonic foraminiferans (Figure 4.2). Foraminiferans incorporate different proportions of ^{16}O and ^{18}O according to the temperature of the seawater in which they grow: the lower the temperature, the higher the $^{18}O/^{16}O$ ratio in the calcium carbonate secreted. Although all organisms secreting calcium carbonate produce $CaCO_3$ with higher $^{18}O/^{16}O$ in cold than warm water, the actual ratio for a particular temperature depends on the species concerned (see Figure 4.2) (and also on the $^{18}O/^{16}O$ ratio of the water it is living in).

The $^{18}O/^{16}O$ ratio of remains of foraminiferans is also affected by the waxing and waning of ice-caps. Water that evaporates from the ocean eventually condenses as cloud and falls as rain or snow. When seawater evaporates, water molecules with the lighter oxygen isotope ($H_2^{16}O$) evaporate more readily, so atmospheric water vapour is relatively enriched in the lighter isotope. When water vapour condenses and is precipitated back into the ocean, the heavier isotope ($H_2^{18}O$) condenses preferentially. Both processes deplete water vapour in the atmosphere in $H_2^{18}O$ relative to $H_2^{16}O$. When ^{18}O-depleted water vapour is precipitated as snow at high latitudes, the snow will also be depleted in ^{18}O relative to the oceans. The larger the ice-caps, the higher the relative proportion of ^{18}O in seawater and the lower the relative proportion of ^{18}O in ice-caps.

Under what conditions, then, will planktonic foraminiferans have high $^{18}O/^{16}O$ ratios?

$^{18}O/^{16}O$ ratios will be high if the surface seawater is cold, and the amount of water held in ice-caps is large.

The amount of ^{18}O in foraminiferan tests is very small, but it can be measured accurately by mass spectrometry. The result is usually reported not as a simple ratio, but as a delta (δ) value, which is determined by comparison of the sample with a standard, and results in a value expressed in parts per thousand (‰ or 'per mil'):

$$\delta^{18}O = \frac{(^{18}O/^{16}O)_{sample} - (^{18}O/^{16}O)_{standard}}{(^{18}O/^{16}O)_{standard}} \times 1000 \qquad (4.1)$$

Generally, the standard used nowadays is seawater; Standard Mean Ocean Water (SMOW), or Vienna Standard Mean Ocean Water (VSMOW), which superseded SMOW in 1995.

QUESTION 4.4 (a) A foraminiferan test has a $\delta^{18}O$ value of $-1‰$ relative to VSMOW. By reference to Equation 4.1, explain whether it contains more or less ^{16}O than the VSMOW standard. (b) What is the meaning of a $\delta^{18}O$ value of $0‰$?

To reconstruct past variations in sea-surface temperature, the 'ice volume' signal must be separated from the temperature signal. One way to do this is to make oxygen-isotope measurements for tests of benthic foraminiferans, since waters near the ocean floor generally remain relatively unaffected by global changes in surface temperature, but are always affected by ice volume changes. By subtracting the $\delta^{18}O$ value for a benthic foraminiferan from the $\delta^{18}O$ value of a contemporaneous planktonic foraminiferan, an estimate of SST at the time in question can be obtained (Figure 4.11).

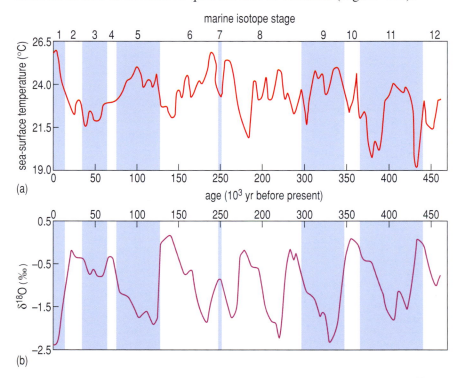

Figure 4.11 (a) Variations in sea-surface temperature in the Caribbean inferred from oxygen-isotope ratios measured for the tests of planktonic and benthic foraminiferans. The oxygen-isotope record for planktonic foraminiferans (b) is also shown. Interglacial (blue) and glacial (white) periods are identified by odd- and even-numbered Marine Isotope Stages, respectively. (See text for details.)

The last glacial period can be seen recorded in Figure 4.11 as higher $\delta^{18}O$ values and lower sea-surface temperatures from ~30 000 to ~12 000 years ago. The isotope record in fact reveals a number of glacial and interglacial periods; these are known as **Marine Isotope Stages** (MIS), and by convention, odd-numbered stages represent interglacials and even numbers glacials. Glacial stages tend to be longer than interglacials, and their termination is rapid (on a geological timescale). Furthermore, the record shows small perturbations in the average climate during both glacials and interglacials (these perturbations are called *stadials* and *interstadials*, respectively).

Figure 4.11 shows that there have been six major glacial/interglacial cycles during approximately the past 0.5 Ma. Since 0.71 Ma ago, there have been eight such cycles.

What is the approximate time period for these climate cycles?

It is close to 100 000 years. This 100 000-year periodicity is less obvious in records of earlier climatic fluctuations preserved in sediments, which, back to about 2.6 Ma ago, are dominated by cycles lasting *c.* 40 000 years.

What processes are seen as controlling the periodicity of these climatic cycles?

They are the cyclical changes in the configuration of the Earth in its orbit, which result in periodic changes in the amount of solar radiation reaching the Northern Hemisphere. These cyclical changes are known as **Milankovitch cycles**; there are three different cycles, superimposed on one another, having periods of about 22 000 years, 40 000 years and 110 000 years.

These orbital variations *alone*, however, cannot explain the causes of glacial periods. One reason is that the pattern of Milankovitch orbital variations, as recognized in oceanic oxygen-isotope records from the Quaternary, must have operated throughout most, if not all, of the Earth's geological history. For much of that time, there is no evidence of any significant glaciation. Another reason is that Milankovitch theory relates the pattern of glacial and interglacial stages to changes in the amount of solar radiation reaching the Northern Hemisphere, but analysis of sediment cores from sites throughout the world ocean reveals that this pattern is in fact global. The mechanism most likely to 'mix and merge' climatic signals in both hemispheres is the pattern of deep ocean circulation and the global thermohaline conveyor.

Other tools used to assess past SST include statistical analysis of the microfossil assemblage (different species tolerate different ranges of water temperature), and measurement of Mg/Ca ratios in shells of well-preserved planktonic foraminiferans (like $\delta^{18}O$, Mg/Ca of a shell varies with temperature, although it does *not* vary with ice volume).

What are the drawbacks of using foraminiferans as palaeoclimate proxies?

Recall that roughly 80% of the ocean floor lies below the CCD. Foraminiferan tests are composed of calcium carbonate and are likely to dissolve below the CCD. This means that records of environmental change based on foraminiferans are often incomplete. An alternative method for obtaining past sea-surface temperatures, this time from *organic* remains of coccolithophores, is described in Box 4.1.

4.3.2 OCEAN CIRCULATION

Along with surface currents, the pattern of thermohaline circulation is an important factor in Earth's climate because its overall effect is to transport vast quantities of heat poleward into high latitudes. The thermohaline circulation involves the formation of North Atlantic Deep Water in the Greenland (and Labrador) Seas, its flow southward in the western Atlantic, round southern Africa (where it is joined by Antarctic Bottom Water) and then into the Indian and Pacific Oceans, and its subsequent return to the surface through upwelling (Figure 2.23). As you read in Chapter 2, this pattern is reflected in the concentration of nutrient elements in bottom seawater: concentrations are highest in the oldest waters in the deep northern Pacific, and lowest in newly formed deep water in the northern Atlantic (Figure 2.22). If deep-water circulation was different in the past, we might expect to see a different distribution of the nutrient elements.

BOX 4.1 MOLECULAR INDICATORS OF ENVIRONMENTAL CHANGE

Alkenones (a type of ketone) are a class of organic molecules found in almost all marine sediments from the present day back to at least 140 Ma ago. They are important constituents of the membranes of coccolithophores, such as *Emiliania huxleyi* (Figure 1.5). It turns out that coccolithophores alter the number of double bonds in their alkenone constituents, depending on the temperature of the water that they are growing in. Careful laboratory experiments show that the relative abundance of alkenones with 37 carbon atoms and two carbon–carbon double bonds increases at higher temperatures, while the relative abundance of alkenones with 37 carbon atoms and three carbon–carbon double bonds increases at lower temperatures. Expressed in simple terms: more double bonds means cooler waters. The so-called **alkenone unsaturation index**, U^K_{37} (where U is for unsaturation, i.e. the presence of double bonds; the superscript K is for ketone; and the subscript 37 is for the number of carbon atoms), is given by:

$$U^K_{37} = \frac{[C_{37:2}]}{[C_{37:2}] + [C_{37:3}]} \qquad (4.2)$$

where $[C_{37:2}]$ and $[C_{37:3}]$ refer to the concentration of alkenones with two and three double bonds, respectively. The relationship between U^K_{37} and temperature (T, °C) is as follows:

$$U^K_{37} = 0.034T + 0.039 \qquad (4.3)$$

QUESTION 4.5 At what time of year would you expect coccolithophore productivity to be highest at temperate latitudes, and what are the implications for measurement of sea-surface temperature using the alkenone unsaturation index?

Figure 4.12 compares records of SST derived from the U^K_{37} index and on the basis of the distribution and relative abundance of different species of planktonic foraminiferans for a core recovered from the eastern North Atlantic. Records derived from the U^K_{37} index are generally in good agreement with those derived from warm (August) foraminiferal assemblages between 28 000 and 8000 years ago, which suggests that maximum coccolithophore production occurred in the summer months in the glacial ocean. The relationship between these two proxies breaks down for sediments deposited during the last 8000 years. This may be linked to a switch in the seasonal timing of maximum coccolithophore productivity from mid-summer in the glacial ocean to late spring—early summer in the modern ocean.

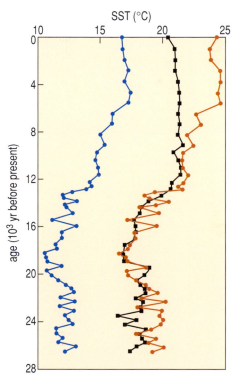

Figure 4.12 A comparison of planktonic foraminiferal SST estimates for August (orange circles, right) and February (blue circles, left), with the U^K_{37} (black squares) palaeotemperature index.

The nutrient elements (principally nitrogen and phosphorus, in the form of nitrate and phosphate) are incorporated into the soft tissues of biogenic material and are therefore not well preserved in sediments. For this reason, we need to find a nutrient proxy; that is, something that mimics the behaviour of the nutrients, and is preserved in the sedimentary record. Cadmium is one such proxy. As with nitrate and phosphate, surface waters have the lowest concentration of dissolved cadmium, and deep waters in the northern Pacific the highest, while the concentration in deep Atlantic waters falls midway between these extremes (Figure 4.13). It turns out that cadmium and calcium are sufficiently similar chemically for cadmium to substitute for calcium in the calcite secreted by foraminiferans. Because of this, the cadmium : calcium ratio in foraminiferan shells should be proportional to the cadmium : calcium ratio in the seawater in which the shell formed. Since the concentration of Ca hardly varies in seawater, and Ca has a residence time in excess of a million years, then any change in the Cd/Ca ratio in seawater over the past 10^6 years or so is likely to result from the redistribution of Cd within the ocean.

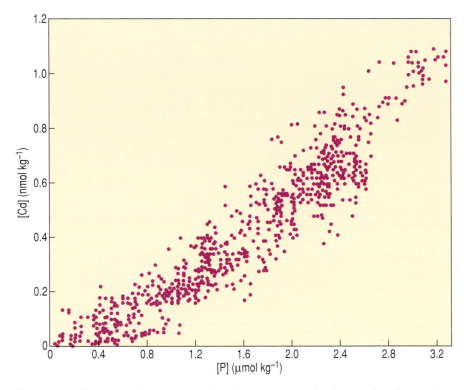

Figure 4.13   Dissolved cadmium versus phosphorus concentration for samples of seawater collected from sites throughout the oceans.

QUESTION 4.6   Explain which end of the line defined by the data points in Figure 4.13 corresponds to surface seawater, and which to deep northern Pacific water.

A number of studies have investigated changes in benthic foraminiferal Cd/Ca over the past 400 000 years. These show that Cd/Ca in benthic foraminiferans from sediments deposited in the North Atlantic during glacials is some 22% higher, indicating that deep water in the North Atlantic must have had higher concentrations of the nutrient elements in glacial times relative to today. One explanation for these higher nutrient levels is that the flux of nutrient-poor NADW was reduced (Section 2.4), allowing nutrient-rich waters from the south (including AABW) to penetrate further north. Measurements of carbon isotopes in benthic foraminiferan shells (see Box 4.2) support this idea.

BOX 4.2   CARBON ISOTOPES AND THE SEDIMENTARY RECORD

Carbon occurs in nature as a mixture of two stable isotopes, carbon-12 (^{12}C) and the much rarer carbon-13 (^{13}C). All the common photosynthetic pathways discriminate against ^{13}C in favour of ^{12}C, so organic matter produced by photosynthesis is enriched in ^{12}C and depleted in ^{13}C relative to the inorganic carbon in the atmosphere and hydrosphere (CO_2 gas, plus hydrogen carbonate and carbonate ions in solution). A typical profile of the ratio of ^{13}C to ^{12}C (expressed as δ^{13}C) of dissolved inorganic carbon in the ocean is shown in Figure 4.14.

How does the shape of the curve relate to the shape of a typical curve for phosphate (Figure 2.10)?

The curve for δ^{13}C is very similar to that for phosphate. This is because photosynthesis leads to depletion of ^{12}C (and phosphate), and therefore relative enrichment of ^{13}C, in surface waters, and respiration returns ^{12}C (and phosphate) to seawater at depth, so deep waters contain relatively less ^{13}C.

Similarly, today's pattern of deep-water circulation means that deep waters in the North Atlantic have relatively high ^{13}C/^{12}C (and low levels of phosphate), while deep northern Pacific waters have relatively low ^{13}C/^{12}C (and high levels of phosphate).

The record of the δ^{13}C value of deep ocean water can be obtained from analysis of the ^{13}C/^{12}C ratio of the shells of benthic foraminiferans. The most reliable records seem to come from species that live *on* the sea-floor (epifaunal species) rather than those that live

within the sediments (infaunal species); the δ^{13}C value of infaunal species tends to be affected by the amount of organic carbon reaching the sea-floor.

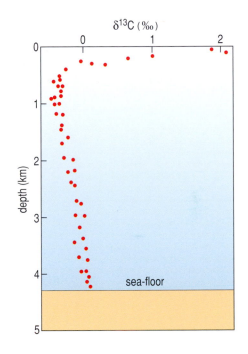

Figure 4.14   Profile of δ^{13}C of dissolved inorganic carbon in the north-western Pacific Ocean. δ^{13}C is calculated from measured carbon-isotope ratios in an analogous way to δ^{18}O:

$$\delta^{13}C = \frac{(^{13}C/^{12}C)_{sample} - (^{13}C/^{12}C)_{standard}}{(^{13}C/^{12}C)_{standard}} \times 1000$$

Having read Box 4.2, would you expect δ^{13}C in benthic foraminiferans to be relatively high or low if they lived in waters that were relatively nutrient-rich?

We would expect the foraminiferans to have relatively low δ^{13}C. Indeed, in glacial times, the δ^{13}C value of benthic foraminiferan shells accumulating on the North Atlantic sea-floor was ~0.4‰ lower than the value for today.

Glacial–interglacial changes in deep-water circulation in the South Atlantic are less clear-cut. Benthic foraminiferal shells recovered from the Southern Ocean show that those deposited at the height of the last glaciation have lower δ^{13}C than those being deposited today. As for the studies in the North Atlantic, this suggests that fluxes of NADW were lower during glacial periods. However, measurements of Cd/Ca in benthic foraminiferans from the Southern Ocean indicate that there was little change in the nutrient concentration of deep water between the

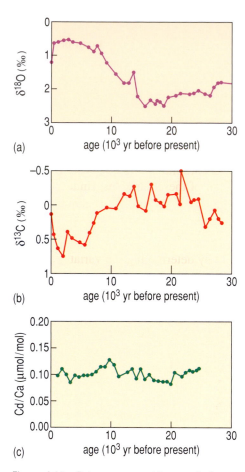

(a)

(b)

(c)

Figure 4.15 Palaeoceanographic records from the Southern Ocean over the past 30 000 years: (a) planktonic foraminiferal $\delta^{18}O$, (b) benthic foraminiferal $\delta^{13}C$ and (c) benthic foraminiferal Cd/Ca. The last glacial period is identified by high $\delta^{18}O$ values and low $\delta^{13}C$ values up to about 15 000 years ago. Meanwhile, Cd/Ca shows little variation.

height of the last glaciation and today (Figure 4.15), which suggests that the flow of North Atlantic Deep Water into the South Atlantic has *not* changed. Why there is this discrepancy between the results from Cd/Ca and $\delta^{13}C$ is not clear. One possible reason is changes in the amount of organic carbon exported from surface waters to the deep sea (which may affect $\delta^{13}C$ in the shells of benthic animals); and another is that the chemical composition of the foraminiferan shells may have changed due to partial dissolution after burial.

The sedimentary carbon-isotope record and gas hydrates

Large shifts (~ −3‰) in the $\delta^{13}C$ value of biogenic carbonates have been recorded in ~55 Ma-old sediments (near the Palaeocene/Eocene boundary), at widely separated locations (Figure 4.16). Other palaeoceanographic studies have shown that around the same time, deep-ocean and high-latitude surface temperatures soared by at least 5–7 °C, and around 35–50% of species of benthic foraminiferans became extinct. The size and the global nature of this $\delta^{13}C$ 'excursion' means that it cannot be attributed to a change in ocean circulation. What, then, was the cause? The consensus view is that the $\delta^{13}C$ excursion was associated with a sudden and massive injection of ^{12}C-rich carbon to the atmosphere or ocean due to decomposition of **gas hydrate** reservoirs in sediments of continental margins; methane from gas hydrates has an average $\delta^{13}C$ value of −60‰. Methane, and the CO_2 released by its oxidation, are both greenhouse gases, so this could have led to sudden warming of the planet.

Proxy studies of periods with high atmospheric CO_2 (such as those in Figure 4.16) are particularly important because on no occasion in the Quaternary have atmospheric CO_2 levels been as high as the present day (370 p.p.m.). If we want to predict climates of the future, with atmospheric CO_2 potentially reaching 2000 p.p.m. when all of the world's coal supplies have been used up, we need to evaluate how the climate system responds to high CO_2 concentrations in the atmosphere and oceans. Palaeoceanographic studies of ancient sediments are the best way of doing this.

Figure 4.16 Records of the variation of $\delta^{13}C$ of biogenic carbonates from various sites around the globe: ODP Site 690 (South Atlantic), Site 865 (Equatorial Pacific), Site 1001 (Caribbean) and Site 1051 (North Atlantic). The data have been placed on a common depth scale. The abrupt drop in $\delta^{13}C$ close to 0 m is common to all sites. Chronological studies at Site 1051 suggest that this carbon-isotope excursion began between 54.93 and 54.98 Ma ago and lasted ~150 000 years.

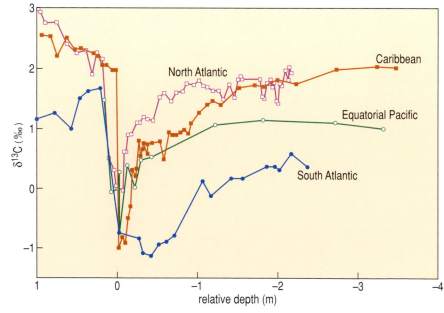

4.3.3 THE CARBONATE SYSTEM

The oceans contain by far the largest reservoir of carbon that can be transferred to and from the atmosphere as CO_2, so any change in the way the marine carbonate system operates has the potential to trigger a change in Earth's climate. As you have already seen, CO_2 does not simply dissolve in seawater but reacts with it to form dissolved CO_2, hydrogen carbonate and carbonate ions (Figure 3.3). The relative proportion of each of these varies with the pH of seawater (Figure 3.4). Meanwhile, the carbonate system is perturbed by biological processes, including photosynthesis and respiration, and the formation of skeletal carbonate. The operation of the carbon cycle in seawater is affected by all of these variables; finding suitable proxies for them in the sedimentary record, however, is not an easy task.

One way to track changes in the carbonate ion concentration of deep water is to reconstruct past depths of the CCD by determining the variation in the calcium carbonate content of sediment cores. As you saw in Section 3.2.1, the tendency for calcium carbonate dissolution is dictated by deep water $[CO_3^{2-}]$; $CaCO_3$ dissolves if seawater is undersaturated with respect to CO_3^{2-}. If the carbonate content of the sediments is 20% or less, then the overlying waters must be below the depth of the CCD.

Firm evidence exists for glacial to interglacial changes in the depth of the CCD. Cores from the tropical Pacific and Indian Ocean reveal that during glacial times the CCD shifted to greater depth. For the Atlantic, the change was in the opposite direction, and dissolution was more intense during glacial times. As the Pacific and Indian Oceans together cover a much larger area than the Atlantic, on a global scale the average CO_3^{2-} concentration of deep water generally appears to have been somewhat higher during glacial times.

What does this mean for levels of dissolved, and therefore atmospheric, CO_2?

Higher $[CO_3^{2-}]$ means lower dissolved $[CO_2]$ and, in turn, lower atmospheric concentrations of CO_2. Ice core data show that levels of atmospheric CO_2 were indeed lower during glacial times, by some 90 p.p.m. (Figure 4.8). So the global average of the change in the depth of the CCD is in the right direction, but it turns out that it can only account for ~20% of the measured change in the concentration of atmospheric CO_2.

The depth of the CCD may, however, be a misleading index of deep ocean $[CO_3^{2-}]$. Release of CO_2 by bacterial respiration of organic carbon in sediments may react with $CaCO_3$ in shells to produce hydrogen carbonate ions:

$$H_2O + CO_2(aq) + CaCO_3(s) \rightarrow Ca^{2+}(aq) + HCO_3^-(aq)$$

The result is that sea-floor dissolution of calcite is to some extent driven by the amount of CO_2 released by respiration into sediment pore waters, as well as by bottom water $[CO_3^{2-}]$. Proxies that are thought to be sensitive to bottom water $[CO_3^{2-}]$ alone have more recently been developed, such as the amount of foraminiferal fragmentation (see Question 4.7), and variations in the weight of foraminiferal tests of specified species and dimensions. These techniques are still being researched but, so far, the perfect proxy for deep water $[CO_3^{2-}]$ remains elusive.

QUESTION 4.7 Figure 4.17 shows palaeoceanographic records for Marine Isotope Stages 12 and 11 for a core from the South Pacific.
(a) First, without looking at Figure 4.17, can you say which MIS (i.e. 11 or 12) corresponds to a glacial period, and which to an interglacial?
(b) Which part of the benthic $\delta^{18}O$ record (i.e. the left-hand or right-hand side) corresponds to MIS 11, and which to MIS 12, and how can you tell?
(c) Use Figure 4.17 to help you decide whether foraminiferans are better preserved during glacials or interglacials. Does your answer agree with the results of the CCD study in the tropical Pacific described above?

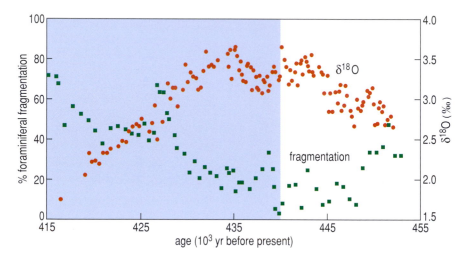

Figure 4.17 Records of benthic foraminiferal $\delta^{18}O$ and % foraminiferal fragmentation from a core in the South Pacific. The percentage of foraminiferal fragmentation is a guide to foraminiferal preservation: a high percentage of fragments (i.e. few whole tests) is indicative of poor preservation (more dissolution), and *vice versa*. Different backgrounds correspond to different marine-isotope stages.

So, if the deeper CCD accounts for ~20% of the reduction in atmospheric concentrations of CO_2 in glacial times, what accounts for the remaining ~80%? Photosynthesis 'consumes' CO_2, so an increase in global biological productivity could also lower atmospheric CO_2, as long as the percentage of organic matter that is preserved on the sea-floor remains the same, or even increases. Therefore, if primary productivity increased in glacial times, there would be increased drawdown of atmospheric CO_2. One way to determine past levels of primary production is to evaluate past levels of nutrient utilization in surface waters.

As we have seen already, the Cd/Ca ratio of benthic foraminiferans can be used as a proxy for deep-water phosphate concentrations. In the same way, Cd in the shells of planktonic foraminiferans can be used as a proxy for phosphate in surface seawater; relatively low levels of Cd (phosphate) mean relatively high nutrient utilization, and therefore high rates of biological production. There are complications with the application of this technique, however. First, in the modern ocean, the ratio Cd/P in surface waters increases from low to high latitudes, and we do not know if it always did. Secondly, uptake of Cd into foraminiferal tests varies with sea-surface temperature, so records of variability in SST are required for accurate interpretation of foraminiferal Cd/Ca. Nevertheless, careful reconstructions of surface water phosphate from planktonic foraminiferal Cd/Ca in sediments from the sea-bed of the Southern Ocean suggest that north of the modern-day Antarctic Front (which is located at ~50° S) there was a similar nutrient utilization during glacial times relative to today (Figure 4.18), which is consistent with what we see for other nutrient tracers (such as the nitrogen-isotope ratio of organic matter: see Question 4.8). South of the Antarctic Front, planktonic foraminiferal Cd/Ca indicates that nutrient utilization was actually lower than today during

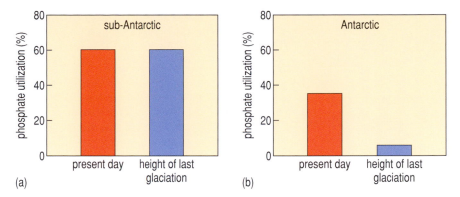

Figure 4.18 Contrasted utilization of phosphate in surface waters at the height of the last glaciation and the present day in the Southern Ocean, based on planktonic foraminiferal Cd/Ca ratios. (a) Sub-Antarctic waters (north of ~50° S); (b) Antarctic waters (south of ~50° S).

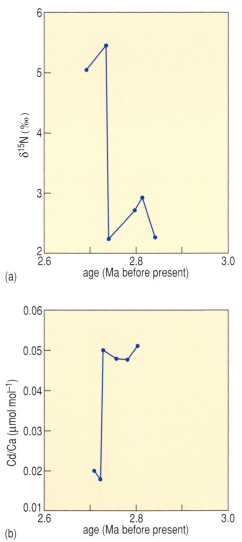

(a)

(b)

Figure 4.19 (a) Nitrogen-isotope and (b) planktonic foraminiferal Cd/Ca data for sediments recovered from a station in the sub-Arctic Pacific, for use with Question 4.8. Note that in (a) $\delta^{15}N$ (‰) = (($^{15}N/^{14}N)_{sample}$ / $(^{15}N/^{14}N)_{standard}$) − 1) × 1000, where atmospheric nitrogen is the standard.

glacial times (Figure 4.18). This is consistent with some other nutrient proxies (including the rate of accumulation of biogenic silica on the sea-floor), but is at odds with results of analyses of nitrogen-isotope ratios of organic matter, which suggest that nutrient utilization was greater in glacial times. The reason for the discrepancy isn't clear at present, but it may be related (in a complicated way which we won't go into here!) to the expansion in sea-ice cover, which, in glacial times, may have been as much as double the surface area of winter ice today, in the region to the south of the modern Antarctic Front.

QUESTION 4.8 Nitrogen isotopes offer an insight into the utilization of nitrate in primary production. The dominant isotope is ^{14}N, and this lighter isotope is preferentially taken up ('fixed') by photosynthesizing organisms, leaving behind the much less abundant heavier isotope, ^{15}N. When nitrate is plentiful, therefore, the $\delta^{15}N$ value (analogous to $\delta^{13}C$ and $\delta^{18}O$) of sea-floor sediments is low, because the primary producers have selected chiefly the lighter ^{14}N isotope. Where nitrate is in short supply, however, proportionally much more of it gets used, including the component with the heavier ^{15}N isotope. So the $\delta^{15}N$ value increases.

(a) The dots in Figure 4.19(a) are $\delta^{15}N$ values in sediments from the sub-Arctic Pacific (close to 50° N, 167° W). Explain how these nitrogen-isotope data are consistent with an abrupt reduction in nutrient upwelling in the northern Pacific gyres 2.73 Ma ago.

(b) Is the planktonic foraminiferal Cd/Ca record shown in Figure 4.19(b) consistent with your answer to part (a)?

4.4 SUMMARY OF CHAPTER 4

1 The sea-floor sediment record extends back to ~200 Ma ago. Sediments are thinnest closest to spreading axes, where the ocean crust is young, and increase in thickness with distance (and depth) from the axis as the crust ages and more time has elapsed for sediments to accumulate.

2 Sediments recovered in deep-sea drill cores contain a wealth of information about past ocean environments. This is obtained from the nature of the sediments themselves, and from measurements of sediment properties (such as the isotopic composition of microfossil shells) which respond systematically to important environmental variables, such as past sea-surface temperature, which cannot be directly measured. Such sediment properties are known as 'proxies'.

3   The most widely used proxy for sea-surface temperature is the oxygen-isotope ratio (usually expressed as $\delta^{18}O$) of planktonic foraminiferans. Foraminiferans incorporate proportionally more ^{18}O into their calcium carbonate shells at low temperatures, and more ^{16}O at high temperatures. Planktonic foraminiferal $\delta^{18}O$ is not the perfect proxy, however. Oxygen-isotope ratios are also affected by the $^{18}O/^{16}O$ ratio of the water in which the foraminiferans lived; this varies with ice volume, and hence sea-level. Other proxies that are useful for determining past sea-surface temperatures include planktonic foraminiferal Mg/Ca, and the alkenone unsaturation index (U^K_{37}).

4   Proxies for the concentrations of the nutrient elements in the deep sea can be used to trace changes in the rate of formation of deep water, and the pattern of thermohaline circulation. One such proxy is the Cd concentration in shells of benthic foraminiferans: Cd mimics phosphate in the deep ocean. Another proxy is the carbon-isotope ratio of shells of benthic foraminiferans. Photosynthesizing organisms take up ^{12}C in preference to ^{13}C, so surface waters are relatively depleted in ^{12}C. This ^{12}C is returned to seawater by respiration at depth. Therefore the older the water mass, the more ^{12}C it contains, and the lower its $\delta^{13}C$ value.

5   Reconstructing past changes in the marine carbonate system is central to understanding the causes of climate change. Deep-water carbonate ion concentration can be estimated from past levels of the CCD, and by analysis of microfossil fragmentation.

6   Changes in biological production can be reconstructed from analyses of Cd in the shells of planktonic foraminiferans: concentrations of Cd (phosphate) in surface waters are low if photosynthetic activity is high, so the proportion of Cd incorporated into the shell decreases.

7   While sediment proxies have provided an abundance of information for reconstructing past climate, they may yield conflicting information. There are a number of reasons for this: our understanding of the parameters that control proxy relationships is incomplete; the chemical signature of microfossils may be altered upon burial; and some proxies require better calibration. Palaeoceanographers are working hard to resolve these conflicts, and to develop new proxies.

Now try the following questions to consolidate your understanding of this Chapter.

QUESTION 4.9   Planktonic foraminiferans preserved in sediments aged ~220 000 years have a $\delta^{18}O$ value of −0.6‰. Explain whether sea-surface temperatures were higher or lower at that time than they are today.

QUESTION 4.10   Results of the analysis of alkenones from within sediments from the Mediterranean are presented in Table 4.1.

(a)   Use Equations 4.2 and 4.3 to complete the Table. The temperature of the topmost four samples has been calculated for you; begin by checking that you get the same answers as ours.

(b)   Plot the sea-surface temperature record on Figure 4.20. Is your record consistent with the planktonic foraminiferal oxygen-isotope record shown in Figure 4.15(a)?

Table 4.1 Alkenone data from a Mediterranean sediment core, for use with Question 4.10.

Age (years before present)	$[C_{37:2}]$	$[C_{37:3}]$	U^K_{37}	Temperature (°C)
1000	22	10.1	0.68	18.8
3220	20	10.5	0.66	18.3
4860	25	13.1	0.66	18.3
7130	18.3	9.6	0.66	18.3
9450	21	8		
11 100	18	11.2		
13 200	25	24		
14 900	13	16		
17 200	19	23.5		
19 000	16	19.7		

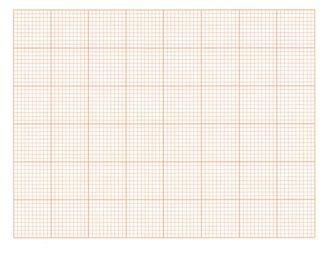

Figure 4.20   Graph paper for answer to Question 4.10(b).

CHAPTER 5

BIOGEOCHEMICAL ACTIVITY IN DEEP-SEA SEDIMENTS

'Below the thunders of the upper deep;
Far, far beneath in the abysmal sea,
His ancient, dreamless, uninvaded sleep,
The Kraken sleepeth ...'

Alfred, Lord Tennyson (1830)

It is tempting to imagine that once sediment has settled on the deep sea-floor beneath several kilometres of water, far removed from waves and storms, it will remain there quite passively until such time as the ocean basins cease to exist. Indeed, that was the consensus view until marine scientists began to take photographs of the deep-sea bed, and, more recently, to explore it using submersibles.

Sediments certainly do not lie undisturbed on the sea-bed after they are deposited. Figure 5.1 and Box 5.1 show that sediments are disturbed by both biological activity and bottom currents. In the extreme, **abyssal storms** have been observed on the sea-floor; these are intermittent erosive episodes, lasting for periods of days to weeks, that tend to occur beneath regions where **mesoscale eddies** develop. The variability and turbulence associated with these eddies can extend to considerable depths, leading to turbulent conditions at the sea-bed. Current speeds can reach more than $0.3 \, m \, s^{-1}$, sufficient to resuspend unconsolidated fine-grained sediment, and to generate ripple marks and other features on the sea-bed (Figures 5.1 and 5.2).

(a)

(b)

(c)

Figure 5.2   Photographs of an abyssal storm event at 4880 m depth off Nova Scotia Rise in the north-west Atlantic. (a) September 30, 1985. The bottom is clean and partially roughened by animal burrows and some tracks. The ridge-shaped ripple feature seen near the bottom centre (just left of the cable) indicates recent bottom activity. The loose cable strand (or rope) extending from the bundle of cables shows the current is flowing from right to left.
(b) October 31, 1985. An intense abyssal storm has smoothed most of the bed, filling in nearly all the burrows and erasing the tracks and other features. The rough area in the lower left is an impact crater formed by a piece of falling debris. The current is now flowing from left to right.
(c) December 27, 1985. Several weeks later, the burrows have been re-opened and there is a general roughening by animal tracks. There has also been some current activity, with 'clouds' of sediment being advected past — some of which has been deposited, smoothing the edges of the impact crater. The current is now flowing from top to bottom of the picture.

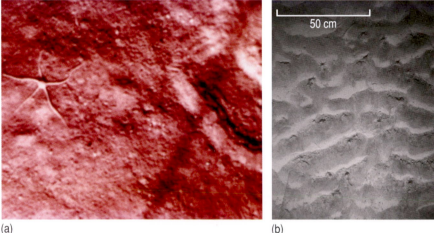

(a) (b)

Figure 5.1   (a) The sea-floor at a depth of 2132 m in the equatorial Atlantic. The animal on the left is a brittle-star of the Echinodermata (Ophiuroidea), which moves over the bottom by 'rowing' — it points one 'arm' forward and pushes itself along using the other four arms. In the upper right is the remnant of a large trail, but the animal that made it was not identified.
(b) Ripple marks in sediments in the Faeroe–Shetland Channel (off the north-west coast of Scotland), at a depth of ~1150 m. Dark sediments are found on the stoss (upstream) face; pale sediments are found on the lee (downstream) face. Organic detritus (dark blobs) can be seen in the troughs of the ripples.

BOX 5.1 CARBONATE MOUNDS

Carbonate mounds are a fascinating example of interplay between sea-floor sediments, bottom currents and biological activity. They tend to form on the upper part of the continental slope at depths between about 500 and 1200 m, and can be as much as 350 m high and 2 km wide at the base, having developed over hundreds of thousands of years (Figure 5.3). The mounds consist of carbonate muds, mainly produced by the breakdown of cold-water corals, such as *Lophelia pertusa* (Figure 5.4), as well as live corals, that colonize the upper and current-facing side of the mound. The mounds and their associated coral reefs also form an important habitat for, amongst others, shrimp, crabs, crinoids and commercial fisheries.

The mounds tend to develop in areas with fast bottom currents and high biological productivity in the overlying surface waters. This gives favourable conditions for coral growth: currents winnow away fine material, exposing gravel, stones and even boulders that provide a hard substrate for coral settlement, while particulate organic material falling from surface layers, and resuspension of bottom sediments by the strong currents, provide food for these suspension feeders.

Figure 5.4   Skeleton of a small colony of the cold-water coral *Lophelia pertusa*. This species is able to grow in water temperatures of 4–12 °C. It grows at rates of around 4 to 15 mm yr^{-1}. In contrast, warm-water corals (e.g. those that form the Great Barrier Reef) cannot tolerate water temperatures lower than 18 °C. They grow at rates of up to about 100 mm yr^{-1}. Image width is ~30 cm.

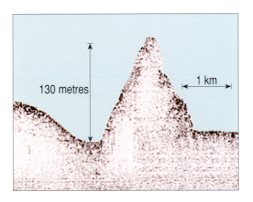

130 metres

1 km

Figure 5.3   Seismic reflection image of a carbonate mound on the West Rockall Bank, off the north-west coast of Northern Ireland.

THE DEEP-OCEAN BENTHIC BOUNDARY LAYER

As a consequence of turbulent motions near the deep ocean floor, there is a well-mixed *benthic boundary layer*, which can be up to several tens of metres thick. The layer affects biological, chemical and geological processes at and near the ocean floor and influences the way these processes 'communicate' with the ocean interior. For example, organic material is biologically transformed within this layer, and its constituents either return to the water column, or enter the sedimentary record.

Particulate material transported in the benthic boundary layer does not only come from vertical settling through the overlying water column; resuspension and lateral advection of particles can also be significant. Lateral transport responds to flow conditions at the sea-bed, bed structure, and benthic activity (e.g. some benthic foraminiferans secrete sticky substances that help to bind sediment particles, making resuspension more difficult). Flow conditions within the benthic

Figure 5.5 Recovery of a benthic lander after deployment in the Arabian Sea. The incubation chamber is located at the very bottom of the instrument; a water-sampling device is located at the top, just above the red battery box.

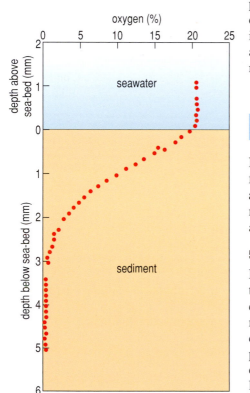

Figure 5.6 High-resolution *in situ* profile of dissolved oxygen in pore waters in sediments at a water depth of 140 m on the Pakistan margin. These measurements were made using a microelectrode, driven into the sediment from a benthic lander.

boundary layer are also important for the dispersal (and settlement) of benthic species.

Measurements of the exchange of material between the sea-floor and the overlying water column, i.e. within the benthic boundary layer, are crucial for quantifying biogeochemical cycles. Investigations of the benthic interface have been sparse until quite recently, mainly because of the difficulty in collecting undisturbed material from the sea-bed. With the development of remotely deployed instrument platforms — 'benthic landers' — it is now possible to examine these processes under *in situ* conditions (Figure 5.5).

Benthic landers operate in two different ways. First, they can insert sensors into the sediments to provide high-resolution profiles of pore water constituents, such as oxygen (Figure 5.6). Secondly, they can partially insert a box into the sediments to enclose overlying waters, forming an *in situ* 'incubation' chamber. The waters are gently stirred and water samples can be collected at pre-determined intervals, whilst electrodes monitor the build-up of dissolved constituents produced by the sediments. All kinds of experiments can be carried out in the chamber: for example, scientists can place different sizes of fluorescent particles (of different colours) into the chamber to assess the extent and rates of sediment mixing and of particle ingestion by benthic animals, and they can study benthic carbon cycling by adding ^{13}C-labelled algal organic matter to the chamber, and tracking its movement into the overlying waters, pore waters and benthic fauna.

5.1 AUTHIGENESIS IN DEEP-SEA SEDIMENTS

New mineral phases formed on the sea-bed, either by direct precipitation from seawater or by the alteration of pre-existing material, are termed **authigenic** deposits. They include ferromanganese deposits, phosphorites, marine barite, hydrothermal deposits such as iron oxides and sulphides, and authigenic clay minerals.

5.1.1 AUTHIGENIC CLAY MINERALS

Much of the montmorillonite in pelagic clays is authigenic rather than terrigenous. This is because seawater rapidly hydrates, devitrifies and decomposes the glassy volcanic material in the volcanogenic component of marine sediments and in the outer rind of lava pillows. Montmorillonite clay is the product of such reactions. The rest of the montmorillonite in pelagic clays is land-derived (Section 1.1.2) and this terrigenous component is provided by weathering of rocks of basaltic composition on land. There is less of the terrigenous component in the Pacific, where volcanism is widespread and land-derived sediment is trapped in trenches.

It is unlikely that clay minerals of the other three main groups (kaolinite, chlorite, illite; Section 1.1.2) are either produced or significantly altered by authigenic reactions on the sea-bed. Their distribution in pelagic sediments is consistent with what is known of the latitude belts in which the clays

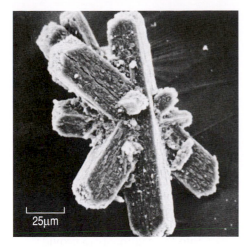

Figure 5.7   Electron micrograph of a cluster of phillipsite crystals recovered from Pacific deep-sea sediments. The formula of phillipsite is $(K,Na,Ca)_3(Al_3Si_5O_{16}).6H_2O$.

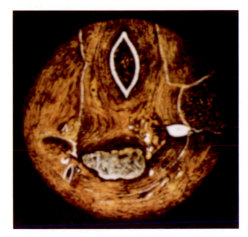

Figure 5.8   Drawing of a section through a small manganese nodule, about 2 cm in diameter. The nodule has grown around a shark's tooth (top centre) and a volcanic fragment (bottom centre), and has incorporated other particles as it grew. This was recovered by the *Challenger* expedition from a depth of 2735 fathoms (*c.* 4350 m).

Figure 5.9   Typical appearance of a rich manganese nodule field in the deep ocean.

originate (kaolinite in low latitudes, chlorite in high latitudes). This would not be the case if these clays were the products of authigenic reactions. Moreover, radiometric dating of clay minerals in pelagic sediments shows that they are relatively old, appropriate to the continental source regions from which the clays probably derived. New phases formed by authigenesis would be younger.

Another authigenic product of reactions between seawater and basaltic glass is *phillipsite*, a potassium-bearing member of a group of hydrated clays called *zeolites* (Figure 5.7). Authigenic minerals such as montmorillonite and phillipsite are formed by the alteration of pre-existing material (basaltic glass); thus, they contrast with newly produced authigenic phases such as iron and manganese oxides, which are the subject of the next Section.

5.1.2   MANGANESE NODULES

Manganese nodules are probably the most widely known of the authigenic deposits found on the deep ocean floor. They are spherical to oblate in shape and range in size from less than 1 cm in diameter to 10 cm or more. Concentrations of nodules on the sea-bed can reach 25 kg m^{-2}.

Manganese nodules were first recovered from the Atlantic Ocean floor, close to the Canary Islands, during the *Challenger* expedition (Figure 5.8). Since then, they have been mapped in all oceans, mainly by photographing the ocean bed. Sometimes their occurrence is patchy, but elsewhere they may form rich nodule fields (Figure 5.9), and in some areas slabs or 'pavements' extend over several square metres and weigh several tonnes. In addition, there are manganese-rich coatings on the surfaces of igneous (usually volcanic) rocks exposed on the ocean floor.

The nodules are mainly mixtures of manganese and iron oxides, and hydroxides which have been deposited in concentric layers around a nucleus of some sort. This may be of biological origin (e.g. a shark's tooth, Figure 5.8), but is more commonly a volcanogenic fragment.

Radiometric dating techniques indicate that deep-sea nodules grow extremely slowly, from a few to a few tens of millimetres per million years. The most important factor affecting nodule growth rate is the rate at which elements are supplied to the deposits.

So how are elements supplied to nodules?

Some nodules obtain elements entirely from seawater (**hydrogenous** deposits), some nodules obtain elements from sediment pore waters (**diagenetic** deposits, Section 5.2), and some obtain elements from both sources (Figure 5.10). Hydrogenous nodules tend to have a smooth surface, while diagenetic nodules tend to have a rough surface.

There are considerable variations in the compositions of nodules from different regions, but all deep-sea nodules show large enrichments of many elements relative to their normal concentration in the Earth's crust. Some elements, such as manganese and cobalt, are concentrated 100-fold or more, and lead and nickel are concentrated about 50- to 100-fold. The metal content of a nodule also depends on how it formed (Figure 5.10); diagenetic deposits are particularly enriched in nickel and copper, up to a maximum of about 3% of Ni and Cu combined, while hydrogenous deposits contain abundant cobalt.

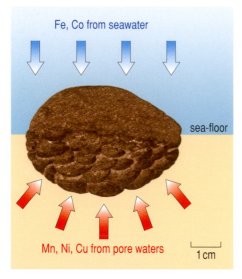

Figure 5.10 Formation of a manganese nodule from both hydrogenous and diagenetic sources. Elements released during diagenesis of marine sediments (see Section 5.2) are supplied to the bottom of the nodule; this usually leads to a rough surface. Elements are supplied to the top of the nodule from seawater, which creates a smooth surface.

The distribution of nodules is a function of a number of different factors. Nodules require a nucleus, and as most nodule nuclei are volcanic in origin, patterns of volcanic activity and the subsequent dispersal of volcanic materials have an important influence on where and in what quantity nodules occur. Nodule production is promoted by high biological productivity in surface waters, because metals are released into sediment pore waters as organic material is broken down. Bottom-dwelling organisms also influence nodule formation — recall that the growth rate of deep-sea nodules is extremely slow (a few millimetres per million years), much slower even than sediment accumulation rates in the deep ocean (a few tens of metres per million years). Foraging benthic organisms help to keep nodules at the sediment surface by constantly nudging them, rolling them over. However, nodules are absent if sedimentation rates are too high, as not even the most frantic activity by benthos could keep a nodule at the sediment surface, so it cannot grow.

QUESTION 5.1 Ferromanganese nodules accumulating on the floor of the semi-enclosed Baltic Sea are unusual in that (1) they grow extremely fast — around 0.02 mm per *year* — but (2) they contain only low concentrations of nickel, cobalt, copper and lead. The nodules are found only at depths shallower than ~50 m; waters below this depth are anoxic.

(a) Calculate the growth rate of nodules in the Baltic Sea in units of mm per million years. How does your answer compare with the growth rates of nodules in the deep sea?

(b) Explain why nodules are absent below ~50 m depth.

THE EXPLOITATION OF DEEP-SEA NODULES

The combined concentrations of metals, in particular copper, nickel and cobalt, make deep-sea nodules an attractive proposition for mining companies. There are three major obstacles to exploitation:

1   The technological problems and cost of extracting a thin layer of nodules from the sea-bed at depths of about 5 km are formidable. One possible method of extracting nodules involves simply sucking them off the bed with a giant 'vacuum cleaner' device.

2   Major environmental disruption would result not only from the dredging or 'vacuuming' of large tracts of the sea-floor, but also from the disposal of the large volumes of muddy 'waste' sediment sucked up along with the nodules.

3   Who owns the high seas? The short answer is nobody, and there is no provision under the Law of the Sea for establishing mineral rights to the sea-bed beneath the open oceans. Deep-sea mining operations are authorized and monitored by the International Sea-Bed Authority, with the aim that no single nation, consortium or operator obtains an unfair economic advantage.

Despite these obstacles, India is forging ahead with plans to mine nodule reserves worth an estimated US$200 billion from a 150 000 km^2 area of the Indian Ocean. The outlook for nodule excavation in the remainder of the deep ocean is rather unclear, being largely dependent on the prices of metals in world markets.

5.2 DIAGENESIS

Diagenesis is a term applied to the chemical changes that occur within sediments, through interaction with pore waters, as they become compacted after burial and eventually lithified and recrystallized to form sedimentary rocks. Diagenetic changes can also affect volcanic rocks when they become buried by sediments.

It is difficult to draw hard and fast lines between authigenic and diagenetic reactions. Here, we define authigenic reactions as those occurring at the sediment–seawater interface (Section 5.1), and diagenetic reactions as those occurring beneath the interface. However, authigenic processes initiated at the sediment–seawater interface commonly continue after burial, as reactions between sediments and pore waters — when they should strictly be termed diagenetic, even though they may still involve the formation or alteration of clay minerals and zeolites. So, we cannot always be sure just when an authigenic mineral phase becomes a diagenetic one, and you may find these terms used in apparently contradictory ways in the literature. Once we get more than a few tens of centimetres below the sediment–seawater interface, the rate of exchange between seawater and interstitial pore waters in the sediments decreases considerably. Reactions between sediment particles and trapped pore waters can then be reliably defined as diagenetic.

During diagenesis, elements can be mobilized into solution and so can migrate through the pore waters (Box 5.2). Some of these elements (together with those already present in the pore waters) are incorporated into newly formed or altered minerals. However, a fraction of the elements can escape capture in this way, and so be released to the overlying seawater.

One of the better known and more obvious diagenetic changes is the gradual long-term lithification of calcium carbonate sediments to form chalk or limestone. The recrystallization commonly involves some replacement of calcium by magnesium to form the mineral *dolomite*, $(Ca,Mg)CO_3$: magnesium is removed from solution in the pore waters while calcium goes into solution.

Siliceous biogenic sediments change in structure and composition even during shallow burial, resulting in loss of some silica to pore waters, and are gradually lithified to form a very hard rock called *chert*. (Varying amounts of silica also occur as quartz in sands and silts in almost all sediments, but this undergoes very little change during diagenesis, beyond some solution and recrystallization along grain boundaries.)

For pelagic clays, chemical changes are very slow, so long as the environment remains oxidizing. This is not really surprising, because the environment at the sea-floor is not very different from that in which continental weathering produces detrital minerals in the first place. Temperatures are generally lower and pressures a good deal higher, but the chemical environment is still oxidizing, even though the minerals are in contact with seawater instead of the atmosphere.

Where anoxic conditions develop within sediments, the chemical environment rapidly becomes quite different from that of continental weathering. This is the subject of the next Section.

BOX 5.2 TRANSPORT OF DISSOLVED CONSTITUENTS IN SEDIMENT PORE WATERS

As particles settle onto the sediment surface, the underlying sediments are compacted, and some pore water is expelled. This upward **advection** of water can transport significant amounts of dissolved constituents. Dissolved constituents can also move through the pore waters as a result of molecular diffusion. This process causes the transport of solutes from regions of higher to lower concentration.

QUESTION 5.2   Figure 5.11 shows a profile for dissolved uranium in pore water in sediments from the San Clemente Basin, which is located off the coast of southern California. How can you tell that uranium is diffusing into the sediments?

Movement of dissolved constituents through pore waters can thus be modelled in terms of advection and diffusion. If the solute undergoes any chemical changes (i.e. it does not behave as a conservative constituent), then a so-called 'reaction' term must also be considered. Although we shall not delve into the details of such models here, some of the types of pore water profiles that result from advection, diffusion and reaction processes are illustrated in Figure 5.12.

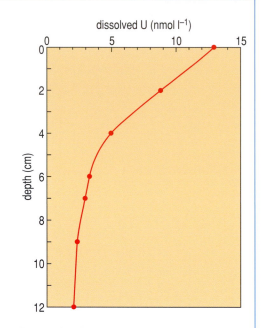

Figure 5.11   Profile for dissolved uranium in pore water in sediments from the San Clemente Basin ($nmol = 10^{-9}\, mol$).

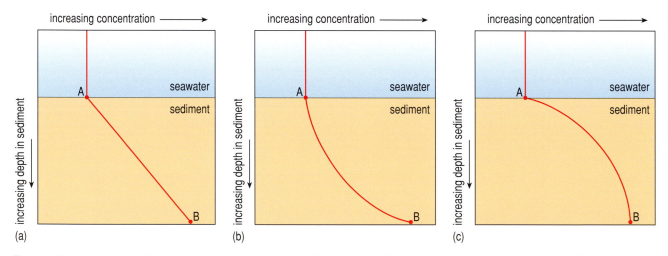

Figure 5.12   Pore water profiles of dissolved constituents. (a) A linear concentration gradient will result if no advection and/or reaction occurs between the sediment–seawater interface (A) and depth B. The solute concentration is controlled by diffusion from a source located at, or below, B into seawater. (b) Negative deviation from the linear concentration gradient results from chemical removal or downward advection of pore water between the sediment–seawater interface and B. (c) Positive deviation from the linear concentration gradient results from chemical production or from upward advection of pore water between B and the sediment–seawater interface.

5.2.1   THE DIAGENETIC SEQUENCE

Most diagenetic chemical changes are driven by redox reactions that involve the oxidation of organic matter. If oxygen is present in sediment pore waters, then organic matter is broken down by *aerobic respiration*, but if the rate of organic matter deposition is fast relative to the rate of respiration by aerobic bacteria, then sub-oxic and anoxic conditions can develop because oxygen in the sediment pore waters cannot be replenished quickly enough by diffusion from overlying seawater. Oxidation of organic carbon is then accomplished through other redox-sensitive constituents found in the sediment and/or pore waters (Table 5.1 and Figure 5.13). In general, different constituents are utilized in succession; constituents which

Table 5.1 Sequence of redox reactions resulting from oxidation of sedimentary organic matter. *Negative values correspond to energy output; the reactions that yield the most energy have the largest negative values.

Process	Oxidant	Product	Energy yield*(kJ mol^{-1})
Aerobic respiration	Dissolved oxygen (O_2)	Carbon dioxide (CO_2). (Also dissolved ammonia (NH_4^+), dissolved nitrite (NO_2^-) and dissolved nitrate (NO_3^-), due to oxidation of organic nitrogen.)	−3190
Manganese(IV) reduction	Manganese oxide (MnO_2)	Dissolved manganese (Mn^{2+})	−2920 to −3090 (depending on mineral type)
Denitrification	Dissolved nitrate (NO_3^-)	Nitrogen (N_2)	−3030
Iron(III) reduction	Iron oxide (Fe_2O_3)	Dissolved iron (Fe^{2+})	−1330 to −1410 (depending on mineral type)
Sulphate reduction	Dissolved sulphate (SO_4^{2-})	Sulphide ions (S^{2-}) (may combine with H^+ ions to form hydrogen sulphide, H_2S, or with Fe^{2+} to form pyrite, FeS_2)	−380
Methanogenesis	Acetic acid (CH_3COOH), or carbon dioxide (CO_2)	Methane (CH_4)	−350

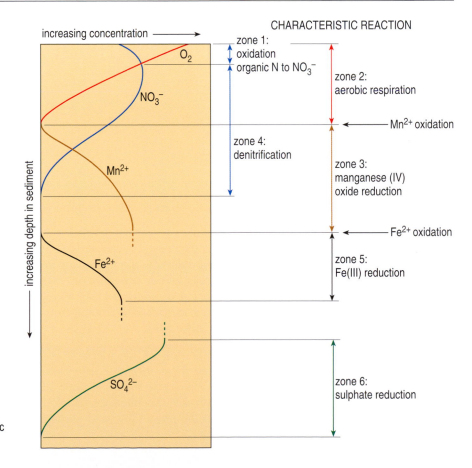

Figure 5.13 Schematic representation of diagenetic zones and trends in pore water profiles during the diagenesis of marine sediments (cf. Table 5.1). Note that nitrate (NO_3^-) increases in zone 1; as oxygen is consumed during the decomposition of organic matter, organic nitrogen is oxidized to nitrate.

take part in reactions that release the most energy are utilized first — diagenesis obeys the laws of thermodynamics. The reactions listed in Table 5.1 generally proceed very slowly. They can, however, be significantly speeded up ('catalysed') by bacteria living within the sediments (see Box 5.3).

QUESTION 5.3 Explain whether bacteria are more likely to utilize dissolved nitrate or acetic acid to oxidize organic matter, given that both are abundant in pore waters.

In pelagic sediments, the sediment accumulation rate is generally very low and the surface oxic layer can extend downwards for several tens of metres. However, in areas of relatively high productivity, some organic matter may escape oxidation in the oxic zone and a diagenetic sequence is initiated in the sediment below. Concentrations of pore water nitrate decrease because of **denitrification**, and concentrations of dissolved Mn^{2+} increase as manganese oxides (usually present as coatings on sediment particles) are reduced (Figure 5.14). The full diagenetic sequence shown in Table 5.1 and Figure 5.13 is rarely seen in deep-sea sediments, because the flux of reactive organic material is generally low.

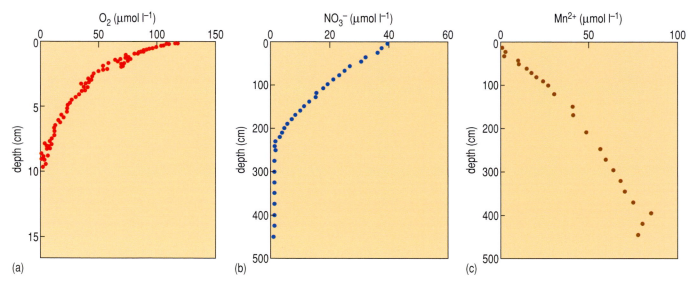

Figure 5.14   Pore water profiles of (a) dissolved oxygen, (b) NO_3^- and (c) Mn^{2+} in the Peru Basin (water depth 4154 m). No dissolved Fe (Fe^{2+}) was detected in these pore waters. Note that the vertical scale for (a) is different from that for (b) and (c).

In the coastal ocean, upwelling areas and other regions of high productivity, the rate of organic matter accumulation is so high that oxygen may penetrate only a few centimetres into the sediment. Organic matter is oxidized using all redox constituents present in the pore waters and sediments, including iron oxides and dissolved sulphate (Table 5.1 and Figure 5.13).

QUESTION 5.4 Figure 5.15 shows corresponding profiles of (a) dissolved and (b) solid-phase manganese. The oxic–anoxic interface is located at ~35 cm below the sea-floor; dissolved Mn^{2+} is transported upwards towards this interface (Box 5.2). Bearing this in mind, explain the shape of the profile for solid-phase Mn.

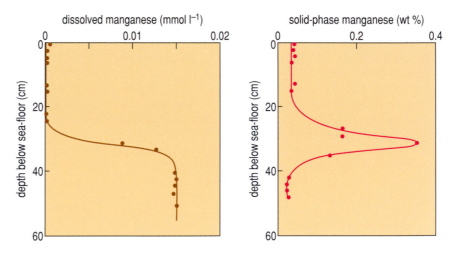

Figure 5.15 Corresponding profiles of (a) dissolved and (b) solid-phase manganese in sediments from the eastern equatorial Atlantic (approximately 0° 31' N, 12° W). At sediment depths <35 cm, the sediments consist mainly of $CaCO_3$ (70–90%), and they contain 0.2–0.5% organic carbon. At sediment depths >35 cm, the sediments are silty, and they contain terrigenous material as well as $CaCO_3$. The organic carbon content of these sediments is 0.5 to >1%.

Redox zones are not always so well defined as implied in Figure 5.13. This is because (1) oxic sediments are disturbed by burrowing benthic animals (**bioturbation**), which can affect the depth of the oxic–anoxic boundary by enhancing the rate of oxygen supply to pore waters. Bioturbation 'mixes up' sediments and pore waters, so the oxic zone tends to be characterized by near-vertical concentration profiles. The lack of oxygen in anoxic sediments makes them unlikely to be bioturbated, and pore waters in the anoxic zone are more likely to have well-defined concentration gradients. Also, as discussed above, (2) the rates of some of the redox reactions listed in Table 5.1 can be very slow, and they can also be catalysed by bacteria. Both factors (1) and (2) can lead to 'smearing' of the different redox zones, so that they grade into one another.

QUESTION 5.5 Oxidation of organic matter by reaction with iron oxides can be represented by:

$$3CH_2O(s) + 2Fe_2O_3(s) \longrightarrow 3CO_2(aq) + 4Fe^{2+}(aq) + 3H_2O$$

(a)   What is the oxidation state of Fe (i.e. Fe(II) or Fe(III)) in (i) Fe_2O_3 and (ii) Fe^{2+}?

(b)   Which element is oxidized, and how can you tell?

You have learnt from previous Chapters that the 'rain' of pelagic sediment to the abyssal plain is continuous and more or less uniform, but can be violently interrupted at irregular intervals of perhaps hundreds to thousands of years by turbidity currents which — even in abyssal plain regions — can deposit as much sediment in a matter of days or weeks as 'normal' pelagic sedimentation supplies in a million years.

How would this affect element mobility during diagenesis?

The turbidity current dumps organic-rich material onto the sea-floor, so the oxic zone moves upwards, closer to the new seawater–sediment interface. Manganese oxides deposited at the 'old' oxic–anoxic interface

BOX 5.3 THE DEEP BIOSPHERE

In 1955, it was declared that the base of the marine biosphere occurs at 7.47 m below the sea-floor as microbiologists were unable to culture bacteria at this or greater depths. Until very recently, it was still believed that bacteria were limited to the upper few tens of metres of ocean sediments, and then quickly died out due to a lack of food sources. Any bacteria found at greater depths were usually considered to be contaminants carried down from above the sediment surface with the sampling equipment. However, detailed studies of bacterial populations (Figure 5.16) and their activities have now been carried out at a number of deep sites, in different oceanographic settings, sampled by the Ocean Drilling Program since 1986.

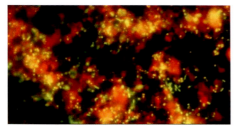

Figure 5.16   Bacteria in sediments recovered from 500 m below the sea-floor. Sediments are stained with acridine orange, and viable bacteria cells can be seen to fluoresce beneath a microscope. This enables microbiologists to estimate bacterial populations in deep-sea sediments.

These studies have shown that bacterial populations are high near the sediment surface and decrease rapidly with increasing depth (Figure 5.17). Even so, the bacterial biomass within the world's ocean sediments is surprisingly large, amounting to about 10% of the total surface biosphere.

The decrease in bacterial populations with increasing depth largely reflects the reduction in degradable organic matter. Although temperature increases during burial (thermal gradients are ~30 °C km^{-1} in the deep ocean), certain types of bacteria (the *hyperthermophiles*) can grow at temperatures up to 121 °C, so temperature does not become limiting to bacteria until several kilometres depth in non-hydrothermal sediments. Sediment porosity also decreases with depth due to compaction, but does not appear to limit bacterial populations to any great extent because bacteria only occupy a very small percentage of the total pore space (about 0.0002%), even at depth.

Bacterial populations can both increase and decrease where environmental conditions change in deeper layers. Bacterial populations at depth within sediments on the Blake Ridge (on the south-eastern margin of North America) are higher than average (i.e. they lie to the right of the solid line on Figure 5.17) because of the presence of gas hydrates (Section 5.2.2), which provide a source of food for bacteria. On the other hand, bacterial populations at relatively shallow depths within sediments at the Juan de Fuca Ridge hydrothermal vent field are lower than average because temperatures within the sediments are unusually high, and cannot be tolerated by the majority of the different species of bacteria.

Various different types of bacteria are now recognized, from cultures isolated from deep sediments, and from analysis of bacterial DNA found within them. These include nitrate reducers, sulphate reducers and methane producers (methanogens). Their distribution usually correlates with geochemical variations; for example, sulphate-reducing bacteria are found where sulphate reduction rates are high and pore water sulphate concentrations are low. This provides direct evidence that bacteriological processes facilitate diagenetic reactions in marine sediments.

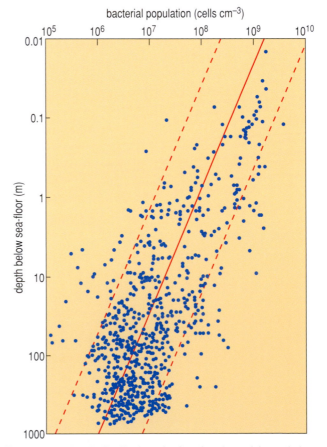

Figure 5.17   Depth distribution of subsurface bacterial populations in the deep ocean. Note that both axes have a logarithmic scale. Data from hydrothermal sites, and sediments containing gas hydrates, have been excluded (see related text). The solid line represents average values; the dashed lines represent the standard deviation (95% confidence limit).

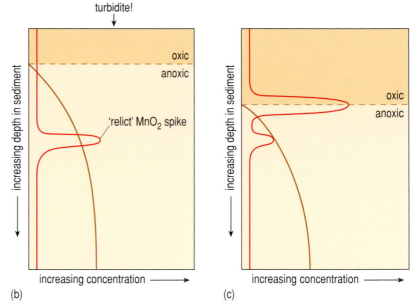

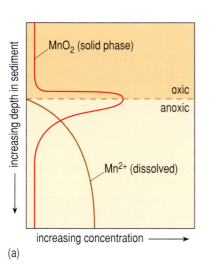

(a)

(b)

(c)

Figure 5.18 Profiles of dissolved and solid-phase manganese in a sediment core from an abyssal plain (a) prior to and (b) just after deposition of a turbidite. If pelagic sedimentation is uninterrupted by a turbidite for long enough (several thousands of years), then the oxic–anoxic interface (denoted by the dashed horizontal line) will have returned to its original depth below the sediment–seawater interface; yet small 'relict' spikes of solid-phase Mn at the depth of the older oxic–anoxic interface can remain, as shown in profile (c).

are reduced to Mn^{2+}, which diffuses upwards and is then oxidized to MnO_2 and redeposited at the 'new' oxic–anoxic interface (Figure 5.18). The manganese oxides are reduced only slowly, however, so 'relict' layers rich in solid phase manganese (and also iron) are common in sediment cores affected by turbidite deposition.

QUESTION 5.6 Sketch the corresponding profile for organic carbon (solid phase) for Figure 5.18(b).

The final reaction listed in Table 5.1 is methanogenesis, which can occur wherever there are large accumulations of organic matter within anoxic sediments. The methane produced can form gas hydrates under appropriate conditions of low temperature and high pressure.

5.2.2 GAS HYDRATES

Gas hydrate is a solid, ice-like substance composed of gas and water molecules. The water molecules form cage-like structures in which various gases, such as methane, ethane, hydrogen sulphide and carbon dioxide, are incorporated to form a *clathrate* structure (Figure 5.19). Gas hydrate occurring naturally in marine sediments is largely composed of methane and water, and is then properly called methane hydrate.

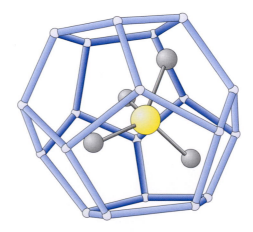

Figure 5.19 Structure of a gas hydrate. A gas hydrate is a crystalline solid in which the molecules consist of a gas molecule (methane, CH_4, is shown here) surrounded by a 'cage' of water molecules.

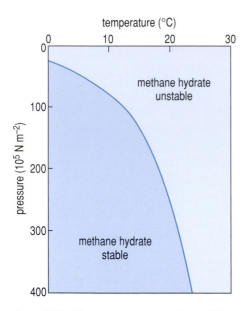

Figure 5.20   Diagram to show how the stability of gas hydrate depends on pressure and temperature. Whether or not gas hydrate actually forms depends on the amount of gas available. (Pressure is measured in newtons per square metre; 10^5 N m^{-2} = 1 bar $\approx$ 1 atmosphere.)

Methane hydrate is stable only at relatively low temperatures and high pressure (Figure 5.20), and can only form if methane gas is available (i.e. below areas of high biological productivity, where organic-rich sediments are preserved by fast burial). These conditions are generally met in sediments along the continental rise (Figure 5.21).

QUESTION 5.7   (a)   The way in which pressure P is related to depth (z) in a column of fluid is described by the **hydrostatic equation**:

$$P = g\rho z$$

where g is the acceleration due to gravity (9.8 m s^{-2}) and ρ is the density. Given that the density of seawater is 1.03×10^3 kg m^{-3}, calculate the pressure due to a 2 km column of seawater. Your answer will be in N m^{-2} (newtons per square metre).

(b)   Use Figure 5.20 to help you to decide whether sediments located at 2 km water depth and (i) at a temperature of 10 °C and (ii) at a temperature of 25 °C, can contain methane hydrate or not.

(c)   Why are gas hydrates generally absent at depths greater than ~1000 m below the sea-floor?

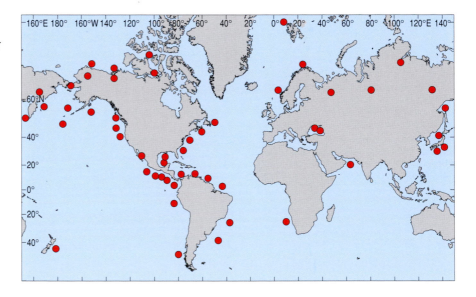

Figure 5.21   Locations of known gas hydrate deposits. As well as along ocean margins, gas hydrates are also found on land, in sediments associated with permafrost at high latitudes.

Gas hydrates can be recognized in drill cores (Figure 5.22), but their presence over large areas can be mapped much more efficiently by acoustic methods, using seismic-reflection profiles. Sound travels much faster (1.8–1.9 km s^{-1}) in gas hydrates than in underlying sediments (1.4–1.6 km s^{-1}), and the sharp drop in velocity at the base of the gas hydrate zone produces a strong reflection called a 'bottom simulating reflection' or BSR (Figure 5.23).

Although gas hydrates have been recognized since the 1970s, interest in them has grown in recent years, for a number of reasons:

1   The worldwide amount of methane in gas hydrates is estimated to contain at least 1×10^{13} tonnes of carbon — about twice the amount of carbon held in all known conventional fossil fuels on Earth. If economically viable techniques are devised to extract this methane, then gas hydrate may become a major energy source.

Figure 5.22   A nearly pure piece of gas hydrate recovered from the Blake Ridge, on the south-eastern margin of North America. The hydrate is white and coated by greenish-grey drilling slurry. Note the bubbles within the slurry; the hydrate is destabilized on recovery from the sea-floor, and soon starts to dissociate into methane gas and liquid water. The core is 10 cm in diameter, and approximately 30 cm long.

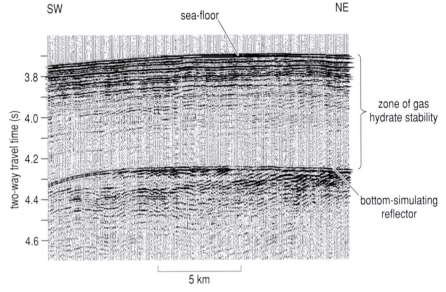

Figure 5.23   Multichannel seismic reflection profile from the crest and eastern flank of the Blake Ridge. The well-defined reflection from the base of the gas hydrate zone roughly parallels the sea-floor (hence the term 'bottom simulating').

2   Gas hydrate is sensitive to changes in temperature and pressure and easily dissociates to release methane to the ocean–atmosphere system (Figure 5.22). Methane is ~20 times as effective a greenhouse gas as carbon dioxide, and gas hydrate may contain three orders of magnitude more methane than exists in the present-day atmosphere. Release of methane from gas hydrates may therefore play a role in global climate change (Section 4.3.2).

3   Gas hydrate apparently cements sediment, and is therefore likely to have a significant effect on sediment strength. Its breakdown is thought to be a cause of submarine landslides.

4   Dissociation of gas hydrate has been put forward as the reason for the supposedly extensive loss of ships in the 'Bermuda Triangle'. While gas hydrates have indeed been detected in this area, and a large gas release could cause a ship to sink by reducing its buoyancy, a check with ships' insurers Lloyds of London reveals that the number of ships that have sunk in this area is not in fact exceptionally high; the claim that the Bermuda Triangle is a potentially dangerous region for ships appears to be a myth!

5.3 SUMMARY OF CHAPTER 5

1   Deep-sea sediments are disturbed and mixed (bioturbated) by animals moving over and through them in search of food. Bottom currents can make bed forms such as ripple marks on the surface, and they can resuspend large amounts of sediment. In regions of the ocean where there are strong western boundary currents, abyssal storms lead to erosion of bottom sediments. In the lowermost few tens of metres, the water column is turbid with suspended sediments and is called the benthic boundary layer.

2   Authigenesis — the formation of new minerals at the sea-bed — includes the formation of clays (montmorillonite and phillipsite) and manganese nodules. The latter are mainly spheroidal structures growing in successive layers around a nucleus at rates of a few millimetres per million years (at least in the deep sea), and reaching average sizes of a few centimetres. They grow by precipitation both from overlying seawater and from pore waters in underlying sediment. Deep-sea nodules contain Co, Ni and Cu in combined concentrations of up to 3%, which make them commercially attractive.

3   Diagenesis encompasses reactions that occur between pore waters and solid phases below the sediment surface, after the sediments become buried. The distinction between authigenesis and diagenesis is not always clear-cut.

4   There are considerable differences in the concentration of redox-sensitive constituents above and below the oxic–anoxic interface in marine sediments. Below the interface, different redox species tend to be utilized in succession: nitrate, manganese and iron oxides, then sulphate. Many of these redox reactions are brought about by bacteria living within the sediments.

5   If the flux of organic carbon to the sediment is extremely high, methane (and other) hydrocarbons form. Under favourable conditions of low temperature and high pressure, large accumulations of gas hydrates may develop.

Now try the following questions to consolidate your understanding of this Chapter.

QUESTION 5.8   Which of the following statements are true, and which are false?

(a)   Transport of dissolved nitrate through pore waters in Figure 5.14(b) occurs only via diffusion.

(b)   Throughout the oceans, sediments become anoxic at depths greater than about 1 m below the sea-floor.

(c)   Manganese is more soluble in reducing than in oxidizing conditions.

(d)   Most material of aeolian origin escapes entrapment in deep-sea trenches.

QUESTION 5.9   Explain whether bacteria living within sediments recently deposited from turbidity currents on the abyssal plain are likely to be more or less abundant than average for a given depth below the sea-floor (cf. Figure 5.17).

QUESTION 5.10 The top 400 m of the sediment sequence collected from the Venezuela Basin of the Caribbean consists of a variety of sediments that go back to the Middle Eocene (about 50 Ma ago). Above the chert band at the base of the sequence (Figure 5.24), the sediments are dominated by biogenic carbonates, but below about 250 m there is a significant siliceous (radiolarian) component. As far as non-biogenic components are concerned, material of volcanic origin is common below about 100 m, where authigenic montmorillonite is the dominant clay mineral; above 100 m, illite and kaolinite are the main clay minerals. There is some evidence for oxygen depletion in the pore waters, but the environment remains oxidizing throughout the sequence.

(a)   Calculate the mean sedimentation rate of the top 400 m of this sediment sequence in mm yr^{-1}.

(b)   Examine Figure 5.24 and determine which of the two constituents represented there become enriched, and which become depleted, in the pore waters of these sediments, with increasing depth.

(c)   From the description of the sediment sequence given above, can you account for the change in the dissolved silica content of the pore waters with depth?

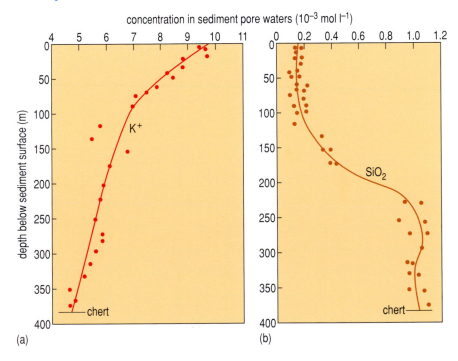

Figure 5.24   Concentration profiles for (a) dissolved potassium and (b) dissolved silica in the pore waters of a sediment core from the Venezuela Basin. Note the layer of diagenetic chert at the base of the sediment sequence.

APPENDIX

PERIODIC TABLE OF ELEMENTS

Group	I	II								transition elements								III	IV	V	VI	VII	0
	1 H 1.01 hydrogen																						2 He 4.00 helium
	3 Li 6.94 lithium	4 Be 9.01 beryllium																5 B 10.8 boron	6 C 12.0 carbon	7 N 14.0 nitrogen	8 O 16.0 oxygen	9 F 19.0 fluorine	10 Ne 20.2 neon
	11 Na 23.0 sodium	12 Mg 24.3 magnesium																13 Al 27.0 aluminium	14 Si 28.1 silicon	15 P 31.0 phosphorus	16 S 32.1 sulphur	17 Cl 35.5 chlorine	18 Ar 39.9 argon
	19 K 39.1 potassium	20 Ca 40.1 calcium	21 Sc 45.0 scandium	22 Ti 47.9 titanium	23 V 50.9 vanadium	24 Cr 52.0 chromium	25 Mn 54.9 manganese	26 Fe 55.8 iron	27 Co 58.9 cobalt	28 Ni 58.7 nickel	29 Cu 63.5 copper	30 Zn 65.4 zinc	31 Ga 69.7 gallium	32 Ge 72.6 germanium	33 As 74.9 arsenic	34 Se 79 selenium	35 Br 79.9 bromine	36 Kr 83.8 krypton
	37 Rb 85.5 rubidium	38 Sr 87.6 strontium	39 Y 88.9 yttrium	40 Zr 91.2 zirconium	41 Nb 92.9 niobium	42 Mo 95.9 molybdenum	43 Tc 98.0 technetium	44 Ru 101 ruthenium	45 Rh 103 rhodium	46 Pd 106 palladium	47 Ag 108 silver	48 Cd 112 cadmium	49 In 115 indium	50 Sn 119 tin	51 Sb 122 antimony	52 Te 128 tellurium	53 I 127 iodine	54 Xe 131 xenon
	55 Cs 133 caesium	56 Ba 137 barium	57 La 139 lanthanum	72 Hf 178 hafnium	73 Ta 181 tantalum	74 W 184 tungsten	75 Re 186 rhenium	76 Os 190 osmium	77 Ir 192 iridium	78 Pt 195 platinum	79 Au 197 gold	80 Hg 201 mercury	81 Tl 204 thallium	82 Pb 207 lead	83 Bi 209 bismuth	84 Po 210 polonium	85 At 210 astatine	86 Rn 222 radon
	87 Fr 223 francium	88 Ra 226 radium	89 Ac 227 actinium	104 Rf 261 rutherfordium	105 Db 262 dubnium	106 Sg 266 seaborgium	107 Bh 264 bohrium	108 Hs 269 hassium	109 Mt 268 meitnerium									

lanthanides

58 Ce 140 cerium	59 Pr 141 praseodymium	60 Nd 144 neodymium	61 Pm 145 promethium	62 Sm 150 samarium	63 Eu 152 europium	64 Gd 157 gadolinium	65 Tb 159 terbium	66 Dy 163 dysprosium	67 Ho 165 holmium	68 Er 167 erbium	69 Tm 169 thulium	70 Yb 173 ytterbium	71 Lu 175 lutetium

actinides

90 Th 232 thorium	91 Pa 231 protactinium	92 U 238 uranium	93 Np 237 neptunium	94 Pu 244 plutonium	95 Am 243 americium	96 Cm 247 curium	97 Bk 247 berkelium	98 Cf 251 californium	99 Es 252 einsteinium	100 Fm 257 fermium	101 Md 258 mendelevium	102 No 259 nobelium	103 Lr 262 lawrencium

Key:
- metals
- metalloids
- non-metals
- all isotopes radioactive

Periodic Table of the elements. Metals occur to the left and non-metals to the right, with semi-metals in between. The named blocks of elements within the Table have their own distinctive properties. The number to the upper left of the chemical symbol is the atomic number of the element; the number to the lower right is its relative atomic mass.

SUGGESTED FURTHER READING

CHESTER, R. (2002) *Marine Geochemistry*, Blackwell Publishers (ISBN 1405101725). A comprehensive and integrated treatment of the chemistry of the oceans, their sediments and biota.

LIBES, S. (1992) *An Introduction to Marine Biogeochemistry*, John Wiley and Sons (ISBN 0471509469). An extremely accessible book that covers all aspects of marine chemistry.

SUMMERHAYES, C. P. AND THORPE, S. A. (1999) *Oceanography: An Illustrated Guide*, Manson Publishing Ltd (ISBN 1874545375). Covers a range of contemporary issues, including marine snow, turbidity currents, ocean chemistry and the marine carbonate system.

WEFER, G. AND FISCHER, G. (1999) *Uses of Proxies in Paleoceanography: Examples from the South Atlantic*, Springer-Verlag (ISBN 3540663401). An advanced text describing research into the reconstruction of past climates, the environment of the South Atlantic, and biogeochemical cycling, by scientists at the University of Bremen.

ANSWERS AND COMMENTS TO QUESTIONS

CHAPTER 1

QUESTION 1.1   Diatoms are algae and depend upon light for photosynthesis. Light intensities are insufficient for net photosynthetic production (i.e. for algal growth) below the photic zone, which rarely extends deeper than about 150 m, and is generally much shallower.

QUESTION 1.2   (a)   The surface-living forms of both foraminiferans and radiolarians are more spindly and delicate, and so are more prone to dissolution than the deeper-water types. Their preservation potential is therefore less.

(b)   Diatoms are primary producers (phytoplankton) whereas radiolarians are secondary consumers (zooplankton). The radiolarians therefore occupy a higher trophic level, so there will be fewer of them both in the water column, and in the sediments.

QUESTION 1.3   Calcareous biogenic sediments predominate along the ocean ridges, which are in general the shallowest regions of the open oceans. Siliceous sediments are most abundant in the Southern Ocean, and occur also along the Equator in the Indian and Pacific Oceans, extending north and south along the eastern Pacific margin. Clays are mostly found in the deepest parts of the oceans, on the abyssal plains. Terrigenous sediments (gravels, sands, silts and clays) form the continental shelf–slope–rise sequences bordering the ocean basins (ice-rafted debris is carried by melting ice from polar regions).

QUESTION 1.4   Sediments are thicker close to continental margins because large quantities of sediment are deposited from rivers on the continental shelf and slope. (Note that a proportion of this sediment can be carried down to the continental rise and the abyssal plains by turbidity currents.) Also, as oceanic crust is formed at mid-ocean ridges and moves away from them, its age increases with distance from the ridge. There is more time for sediments to accumulate, so thicknesses should be greater near continental margins than ridges.

QUESTION 1.5   Yes. Diatoms and radiolarians (Figures 1.8 and 1.9) deplete surface waters of silica to make their skeletons. Dissolution of some of the remains as they sink towards the sea-bed (and after they reach it) enriches the deep waters in silica. As you will see, the effect is much greater for SiO_2 than for Ca^{2+}, because the oceanic reservoir of silica is so much smaller.

QUESTION 1.6   (a)   Kaolinite is produced by chemical weathering in low latitudes, so should be abundant in the equatorial Atlantic.

(b)   Chlorite is characteristic of physical weathering in high latitudes, so should be abundant round Antarctica.

(c)   Illite is a product of continental weathering in general, and there is much more land in the Northern than in the Southern Hemisphere.

CHAPTER 2

QUESTION 2.1   (a)   The profiles for nitrate and barium show that both these constituents are removed from solution in surface water and returned to solution in deeper waters. The straight profile for sodium suggests that there is no mechanism for removing the element from solution in the main body of the oceans; but see (c) below.

(b)   Sodium is (i) conservative; nitrate and barium are (ii) non-conservative.

(c)   Sodium is not *fixed* in organic tissue or skeletal material in the same way as (say) phosphate and silica (and nitrate and barium) are. It is an essential ingredient of body fluids rather than part of the body fabric. Sodium and other conservative constituents are by no means biologically inert, but their biological involvement relative to their concentrations in seawater is such that it does not show up in their concentration profiles.

QUESTION 2.2   If the oceans take 500 years on average for a complete mixing or turnover cycle (i.e. for surface water to be mixed down, or sink to the bottom and come back up again), then a dissolved constituent which spends only 100 years in solution will have been removed long before the cycle is complete, so will not be equally distributed throughout the oceans.

QUESTION 2.3   (a)   Well-stratified surface waters are gravitationally stable. Nutrients are carried down from the surface in sinking organic debris, and cannot be replaced except by slow diffusion processes. If the water column is well-mixed, on the other hand, nutrients returned to deep waters due to decomposition of organic matter may be carried back up to the surface by turbulence.

(b)   Like nitrate and phosphate, silica is dissolved and recycled in upper parts of the water column (Figure 2.10(c)), but it forms skeletal material which is more resistant to solution than soft tissue, and therefore sinks deeper before significant dissolution occurs.

QUESTION 2.4   The profile for cadmium should resemble that for phosphate in Figure 2.10(a). You were not expected to put a scale on your sketch, but for interest and information Figure A1 shows a profile for cadmium in the north-west Indian Ocean.

QUESTION 2.5   (a)   For profile (a), where the concentration decreases with depth from the surface and the location is well away from the influence of major rivers, the source is probably mainly wind-blown dust from the Sahara. For profile (b), where the maximum concentration is at about 3.3 km depth, the only possible source is hydrothermal solutions from the mid-Atlantic ridge axis.

(b)   The hydrothermal source is much the more important. The maximum concentration in profile (b) is more than 20 times greater than that in profile (a).

QUESTION 2.6   (a)   Se(IV) and Se(VI) are clearly being redissolved and recycled in upper parts of the water column. The concentration profile for organic-bound Se resembles that of a scavenged element as its concentration is highest in surface water and decreases with depth.

(b)   Concentrations of total dissolved selenium are lowest in surface waters, and increase with depth (Figure A2). Thus, the concentration profile of total dissolved selenium may be classified as that of a bio-intermediate constituent; however, as we have seen, this masks the fact that concentration profiles of some of the individual forms of dissolved selenium fall into other classes.

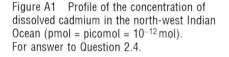

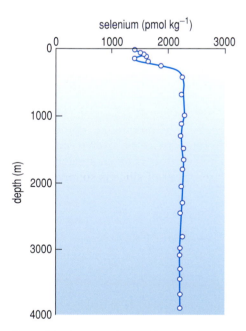

Figure A1   Profile of the concentration of dissolved cadmium in the north-west Indian Ocean (pmol = picomol = 10^{-12} mol). For answer to Question 2.4.

Figure A2   Concentration profile of total dissolved selenium. This is constructed by summing dissolved Se(IV), Se(VI) and organic-bound Se concentrations (Figure 2.19).

(c)   Se(IV) must be taken up in preference to Se(VI) because its profile is close to zero at the surface (i.e. it is biolimiting), whereas that for Se(VI) drops to only about 500 pmol kg^{-1} (i.e. it is bio-intermediate).

QUESTION 2.7   For both (a) and (b), the elements are vertical neighbours in the Periodic Table. Strontium accompanies calcium into calcareous skeletal material (Section 2.1) and lies immediately beneath it in Group II of the Periodic Table (see Appendix). Germanium accompanies silicon into siliceous skeletal material and lies immediately beneath it in Group IV. Note that uptake by analogy occurs only with *some* vertical neighbours. It does not apply to others, such as carbon and silicon.

QUESTION 2.8   Assumption 2 ruled out the conservative constituents, which are not removed from seawater to any significant extent by biological activity. Assumptions 1 and 3 ruled out scavenged constituents, which can have significant aeolian and hydrothermal inputs, and whose rates of input and removal vary considerably with time and from place to place.

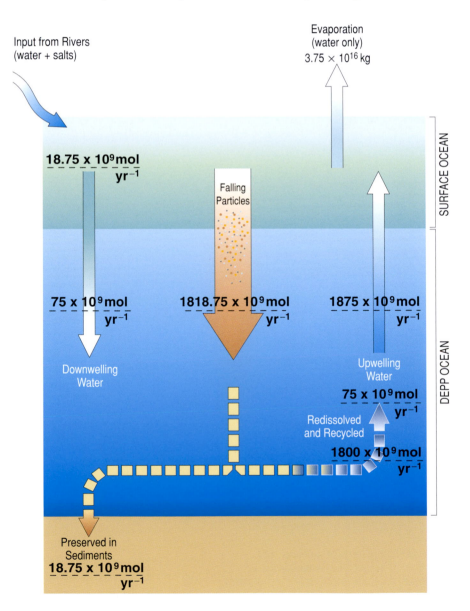

Figure A3   Completed Figure 2.21.

QUESTION 2.9 For 1(a), we need to work out the following fraction as a percentage:

$$\frac{\text{amount in falling particles}}{\text{amount entering surface ocean}}$$

$$=\frac{1818.75 \times 10^9}{(18.75+1875)\times 10^9}\times 100$$

$$=\frac{1818.75\times 10^9}{1893.75\times 10^9}\times 100 = 96\%$$

For 1(b), we need to work out the following fraction as a percentage:

$$\frac{\text{amount preserved in sediments}}{\text{amount entering surface ocean}}$$

$$=\frac{18.75 \times 10^9}{1893.75\times 10^9}\times 100 = 1\%$$

For 2, we need to work out the following fraction as a percentage:

$$\frac{\text{amount preserved in sediments}}{\text{amount in falling particles}}$$

$$=\frac{18.75 \times 10^9}{1818.75\times 10^9}\times 100 = 1\%$$

QUESTION 2.10 There is a striking difference. Nitrogen and phosphorus have the lowest ratios because they form soft tissue and are mainly recycled in surface waters. Next comes the ratio for carbon, which forms both soft tissue and skeletal material (calcium carbonate); so more carbon reaches the deep ocean before it dissolves (bear in mind that much calcium carbonate escapes dissolution, however — Figure 1.12). The largest ratio is for silicon, which forms skeletal material that mostly dissolves in deep water (see Section 3.1.1).

QUESTION 2.11 Profiles for scavenged elements show that concentrations *decrease* with depth. The older the water, therefore, the more time there is for elements in this group to be adsorbed and scavenged — the older the water, the lower the concentration becomes. As deep water of the Atlantic Ocean is, on average, younger than deep water of the Pacific Ocean, concentration ratios (see Figure 2.17) will be <1.

QUESTION 2.12 Argon is an inert gas and nitrogen is almost inert (cf. note 4 to Table 2.1), so both these gases can be considered to behave conservatively. Oxygen does not react with seawater, but is used by organisms in respiration and so its concentration is changed by processes other than mixing; it is non-conservative. Carbon dioxide is also non-conservative, as it reacts with water, and so its concentration is also changed by processes other than mixing.

QUESTION 2.13 (a) Values of w for (i) unreactive gases are generally less than unity, whereas for (ii) reactive gases they may be as high as several thousand. From the definition of w, therefore, rainfall will have much higher concentrations of reactive gases and much lower concentrations of unreactive gases than the surrounding atmosphere.

(b) Values of w can be as high as 10^5 to 10^6 for particles, but are only exceptionally as high as 10^5 for gases. This implies than in general rain scavenges particles more efficiently than gases.

QUESTION 2.14   The profile for pH follows that for oxygen quite closely, indicating that the water is most acid in the oxygen minimum layer. Particulate organic matter tends to accumulate in this layer; oxygen is used up in its decomposition, and CO_2 is produced (reverse of reaction 2.3). This CO_2 will combine with water molecules to form carbonic acid (see Section 3.1.2).

QUESTION 2.15   According to Figure 2.23, the oldest bottom waters are in the northern Indian and Pacific Oceans. We should expect them to be most impoverished in dissolved oxygen, as well as being enriched in nutrients.

QUESTION 2.16   There is more carbon in this ratio because skeletal carbonate = $CaCO_3$, so it is accompanied by 1 mole of C. The 25 moles of Ca must therefore bring with them 25 'extra' moles of carbon, and $(106 + 25) = 131$.

QUESTION 2.17   The data for zinc show depletion in surface water, enrichment at depth, i.e. nutrient-type or recycled behaviour. Enrichment in deep water is greater at station II, so this is probably the Pacific station. We can confirm with the cerium data, which show scavenged behaviour, i.e. depletion in deep water. There is less cerium in the samples from station II, which is consistent with it being in the Pacific (the older the water, the more time there is for particle-active elements to be adsorbed and scavenged).

QUESTION 2.18   Average ocean-wide concentrations for scavenged elements cannot be very meaningful, because several have residence times *less* than the mean oceanic mixing time. The concentration of these elements is not uniform and reflects the influence of more or less localized sources (and sinks).

QUESTION 2.19   The two-box model can be applied to cyclic salts as long as the dissolved constituents concerned fall into the recycled category. If the Input from Rivers is corrected for the cyclic salt contribution, then the two-box model can provide an estimate of the proportion of 'new' material being removed to sediments each year.

QUESTION 2.20   The concentration in surface waters can be used for ΔC in the equation because of the enormous difference between the actual concentrations of DMS in surface water and its concentration in the atmosphere, which is only a few ng $(10^{-9}\,g)\,l^{-1}$. There are enough uncertainties in gas flux calculations for the difference between, say, $300\,ng\,l^{-1}$ and $(300-3)\,ng\,l^{-1}$ to be neglected.

QUESTION 2.21   (a)   False. Most of them are, but carbon and, to a lesser extent, calcium, are two notable exceptions. Their involvement in biological processes is on a sufficiently large scale to affect their concentrations in seawater.

(b)   False. These constituents are heavily involved in biological processes, and their concentrations are thus changed by processes other than mixing.

(c)   False. There cannot be more water coming up than is going down.

(d)   False. Continental shelves are kept oxygenated as waves and tidal currents mix down oxygen from the atmosphere. They are far too shallow to be affected by deep water masses.

(e)   True. In anoxic water, SO_4^{2-} is reduced to sulphide (reaction 2.5). There is less sulphate in solution, so the $SO_4^{2-}:S$ ratio must fall.

CHAPTER 3

QUESTION 3.1   Because the saturation horizon denotes the depth at which dissolution begins, whereas Figure 3.1 denotes the depth at which dissolution is complete.

QUESTION 3.2   (a)   The saturation horizon corresponds to the transition from waters oversaturated to waters undersaturated with respect to calcite (i.e. where $\Delta CO_3^{2-} = 0$). This level occurs where the vertical dashed line crosses the profiles, i.e. at (i) ~4000 m in the Atlantic, and (ii) ~2800 m in the Pacific. The saturation horizon is deeper in the Atlantic than in the Pacific because Pacific waters are CO_2-enriched and therefore CO_3^{2-}-depleted (Equation 3.7); this is due to thermohaline circulation patterns that cause deep Pacific waters to become isolated from the surface for longer (Figure 2.23).

(b)   Profiles for aragonite saturation will have a similar shape to those for calcite saturation, but the depth of the saturation horizon will be shallower. This is because aragonite is more soluble than calcite.

QUESTION 3.3   The glacial sediments, because much of this material is likely to have been transported within ice and so protected from chemical attack.

QUESTION 3.4   The sea-bed is far below the depth at which the profile crosses the aragonite saturation horizon. It is also some way below the calcite saturation horizon. This indicates that there is likely to be substantial dissolution of calcium carbonate.

QUESTION 3.5   Deeper water is relatively cold and under pressure, and relatively rich in ΣCO_2 (e.g. Figure 3.5). The solubility of CO_2 gas in water decreases as (i) the pressure falls, and (ii) upwelled water warms at the surface, and/or mixes with warmer surface water. The reaction shown in Equation 3.3 moves to the left, and the net flux of CO_2 is from sea to air.

QUESTION 3.6   The minimum in the $[CO_3^{2-}]$ profile at about 1000 m depth coincides with the oxygen minimum layer, where oxygen consumption by respiration and bacterial decomposition is at a maximum (Figure 2.27). As respiration and bacterial decomposition produces CO_2, we can expect ΣCO_2 to reach a maximum (and pH a minimum; see Question 2.14) also at this depth; $[CO_3^{2-}]$ is inversely proportional to ΣCO_2.

QUESTION 3.7   (a)   Along Atlantic margins. There are deep ocean trenches around the Pacific, and these trap sediment and prevent the formation of submarine fans.

(b)   Increase. The more rapid the rise, the greater the rate of erosion. The more sediment delivered to the coast, the more accumulates on the continental shelf and the greater the number of turbidity currents.

QUESTION 3.8   False. Most abyssal plain sediments are deposited from turbidity currents.

CHAPTER 4

QUESTION 4.1   (a)   Locality 1 is near the spreading axis (ridge crest), and must lie above the carbonate compensation depth (CCD), so that calcareous sediments are preserved.

(b)   At locality 2, the crust has subsided below the CCD. Calcareous skeletal material is dissolved (mostly at the sea-bed), so the main sedimentary component is clay (this is probably beneath an oceanic region of low productivity, so there is little siliceous debris).

(c)   At locality 3, the sea-floor is beneath the equatorial region of high productivity. Both siliceous and calcareous skeletal material is produced, but as the sea-floor is still beneath the CCD, only siliceous remains are preserved.

QUESTION 4.2   (a)   Turbidites provide poor quality sediment records because they do not reflect conditions in the overlying water column; they consist of material derived from the continental shelf, which may have travelled many thousands of kilometres before settling on the sea-floor. (Luckily, turbidites are relatively easy to recognize in sediment cores (Chapter 3).)

(b)   If the sedimentation rate is low, then the sediment record is of low resolution; a 1 cm thickness may represent a time period of some several thousand years.

QUESTION 4.3   The isotherms indicate that at the height of the glaciation, the warm-water flow of the Gulf Stream was not operating in the same way as it does today, bringing preferential warming to the north-east margin of the North Atlantic.

QUESTION 4.4   (a)   Equation 4.1 tells you that $\delta^{18}O$ is negative when $(^{18}O/^{16}O)_{sample} - (^{18}O/^{16}O)_{standard}$ is negative, i.e. when $(^{18}O/^{16}O)_{standard}$ is bigger than $(^{18}O/^{16}O)_{sample}$. This means that the foraminiferan test has a higher concentration of ^{16}O than the standard (VSMOW in this case).

(b)   A $\delta^{18}O$ value of zero means that the $^{18}O/^{16}O$ ratio in the sample is equal to that in the standard, by definition.

QUESTION 4.5   Coccolithophore productivity at temperate latitudes is likely to be highest in spring. This means that the alkenone unsaturation index will mainly record sea-surface temperatures at this time.

QUESTION 4.6   Surface seawater corresponds to the left-hand end of the line as it has low concentrations of cadmium and phosphorus (because they are used up in biological production). Deep northern Pacific waters correspond to the right-hand end of the line because they are the oldest and so have highest levels of cadmium and phosphorus.

QUESTION 4.7   (a)   By convention, odd-numbered stages represent interglacials and even numbers glacials (Section 4.3.1). Therefore, MIS 12 corresponds to a glacial period, and MIS 11 to an interglacial.

(b) $\delta^{18}O$ values are relatively high when waters are cold and ice volume is large (cf. Figure 4.11). $\delta^{18}O$ values tend to be higher on the right-hand side of Figure 4.17, so this corresponds to MIS 12. Interglacials are characterized by relatively low $\delta^{18}O$ values; these are found on the left-hand side of Figure 4.17, which corresponds to MIS 11.

(c)   The percentage of foraminiferal fragmentation is generally lower during glacial times, indicating better preservation of carbonate material. This is consistent with the results of studies of the $CaCO_3$ content of sediments from the tropical Pacific, which indicate that the CCD was shifted to greater depths during glacial times.

QUESTION 4.8   (a)   The $\delta^{15}N$ values increase sharply at the 2.73 Ma boundary, from around 3‰ to over 5‰. This indicates an increase in the proportion of the heavier ^{15}N isotope in the sediment, implying that more of the available nitrate was being used up, which in turn implies that less nitrate was being supplied to surface waters, i.e. that upwelling had slowed down.

(b)   Yes, because cadmium is a proxy for phosphate and the Cd/Ca ratio decreases as the $\delta^{15}N$ value increases, implying that more phosphate (as well as nitrate) was being used up when upwelling slowed down.

QUESTION 4.9   A $\delta^{18}O$ value of $-0.6‰$ indicates that the foraminiferans are enriched in ^{16}O relative to modern-day seawater (VSMOW). Foraminiferans incorporate proportionally more ^{16}O into their calcite tests if the water they live in is relatively warm, so this suggests that sea-surface temperatures were higher 220 000 years ago than they are today, provided the $\delta^{18}O$ value of the water in which they formed was not significantly different from today. (Note that the temperature and 'ice volume' or 'sea-level' effects operate in the same direction (although they are not related in a linear way).)

QUESTION 4.10   (a)   See completed Table 4.1.

$$U^K_{37} = \frac{[C_{37:2}]}{[C_{37:2}] + [C_{37:3}]}$$

So, for the sample aged 1000 years,

$$U^K_{37} = \frac{22}{22 + 10.1} = 0.68$$

Equation 4.3 states that

$$U^K_{37} = 0.034T + 0.039$$

Re-arranging

$$T = \frac{U^K_{37} - 0.039}{0.034}$$

So, for the sample aged 1000 years,

$$T = \frac{0.68 - 0.039}{0.034} = 18.8\,°C$$

Following the same procedure for the other samples gives:

Table 4.1 completed.

Age (years before present)	[$C_{37:2}$]	[$C_{37:3}$]	U^K_{37}	Temperature (°C)
1000	22	10.1	0.68	18.8
3220	20	10.5	0.66	18.3
4860	25	13.1	0.66	18.3
7130	18.3	9.6	0.66	18.3
9450	21	8	0.72	20.0
11 100	18	11.2	0.62	17.1
13 200	25	24	0.51	13.8
14 900	13	16	0.45	12.1
17 200	19	23.5	0.45	12.1
19 000	16	19.7	0.45	12.1

(b)   The alkenone unsaturation index suggests that sea-surface temperatures were relatively constant between 20 000 and 15 000 years ago (Figure A4). Between 15 000 and 10 000 years ago, SST increased sharply, and has remained relatively constant ever since. The increase in SST approximately coincides with the sharp decrease in planktonic foraminiferal $\delta^{18}O$ at 15 000 years ago shown in Figure 4.15(a). This is what we would expect as lower $\delta^{18}O$ values also indicate higher sea-surface temperatures.

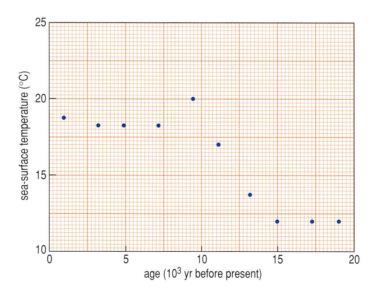

Figure A4   Figure 4.20 completed.

CHAPTER 5

QUESTION 5.1   (a)   The nodules grow at around 0.02 mm per year, so in 1 million years, they grow:

$$0.02 \text{ mm yr}^{-1} \times 10^6 \text{ yr} = 2 \times 10^4 \text{ mm}.$$

This is about 3 orders of magnitude faster than the growth rate of nodules in the deep sea (a few to a few tens of millimetres per million years). (The reason that concentrations of nickel, cobalt, copper and lead in these nodules is low is because they have so little time to accumulate.)

(b)   Both manganese and iron are soluble under reducing conditions, so ferromanganese nodules cannot form in anoxic waters. (Transport of dissolved manganese and iron from anoxic into oxic waters means that the nodules can grow extremely rapidly.)

QUESTION 5.2   Uranium must be diffusing into the sediments because its concentration in the pore waters is less than at the sediment–seawater interface.

QUESTION 5.3   Redox reactions involving dissolved nitrate produce more energy ($\sim3030\,kJ\,mol^{-1}$) than those involving acetic acid ($\sim350\,kJ\,mol^{-1}$), so bacteria are more likely to utilize nitrate.

QUESTION 5.4   Concentrations of manganese in the solid phase are highest at the oxic–anoxic interface as the dissolved Mn^{2+} that diffuses upwards from below the interface is oxidized here. Mn^{4+} is relatively insoluble, so it precipitates (mainly as manganese oxides). Concentrations of Mn in the solid phase are slightly lower beneath the interface than above, as Mn^{4+} is reduced to soluble Mn^{2+} in anoxic (and suboxic) sediments.

QUESTION 5.5   (a) (i)   In Fe_2O_3 there are three oxygen atoms (six negative charges) so the two ions of Fe must be Fe^{3+}, to provide the balancing six positive charges. The Fe is therefore in the Fe(III) oxidation state.
(ii) Fe^{2+} has two positive charges, so it is in the Fe(II) oxidation state.

(b)   The proportion of oxygen is greater in CO_2 than it is in CH_2O, so the carbon is being oxidized. Conversely, oxygen is removed from Fe_2O_3 to form Fe^{2+}, so iron is being reduced.

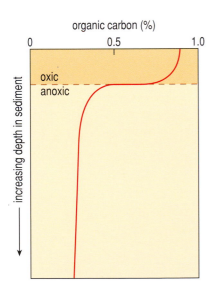

Figure A5   Corresponding concentration profile for organic carbon for Figure 5.18(b). Scale on the horizontal axis is given for information only.

QUESTION 5.6   The concentration profile for organic carbon will look like Figure A5. The turbidite will be relatively rich in organic carbon, as it originates from the upper continental slope, while the surface layer represents 'normal' pelagic sedimentation that is poor in organic carbon.

QUESTION 5.7   (a)   For a 2 km column of water:
$$P = 9.8\,m\,s^{-2} \times 1.03 \times 10^3\,kg\,m^{-3} \times 2000\,m = 2 \times 10^7\,N\,m^{-2}$$

(b)   The pressure at 2 km water depth is $2 \times 10^7\,N\,m^{-2}$ ($200 \times 10^5\,N\,m^{-2}$), so gas hydrate (i) is stable at a temperature of 10 °C (point lies to left of the stability curve), but (ii) it is not stable at a temperature of 25 °C (point lies to right of the stability curve). Note, however, that even if temperature and pressure conditions are favourable, gas hydrate can only form where sufficient gas is available.

(c)   Temperature increases by about 30 °C km^{-1} below the sea-floor (Box 5.3). As gas hydrate is unstable at temperatures greater than ~25 °C (Figure 5.20), it is unlikely to be found at depths greater than ~1 km below the sea-floor.

QUESTION 5.8   (a)   False. The profile shown in Figure 5.14(b) shows a negative deviation from a linear concentration gradient, which indicates that there is either upward advection of pore water, or chemical removal of NO_3^- from solution.

(b)   False. In pelagic sediments, the surface oxic layer can extend downwards for several tens of metres. In the coastal ocean, in upwelling areas and other regions of high productivity, the rate of organic matter accumulation is so high that oxygen may only penetrate a few centimetres into the sediment.

(c)   True. Manganese is more soluble in reducing conditions.

(d)   True. Most aeolian (wind-blown) material is carried far out into the ocean basins, away from continental margins.

QUESTION 5.9   Turbidites usually contain plenty of degradable organic-rich material, so the abundance of bacteria (which feed on this organic material) will be greater than average.

QUESTION 5.10   (a)   400 m (4×10^5 mm) of sediment has been accumulated in 50 Ma (50×10^6 yr), giving an average sedimentation rate of 8×10^{-3} mm yr^{-1}. This is typical for pelagic sediments.

(b)   The profiles show that K$^+$ is depleted in the pore waters of this sequence, whereas SiO$_2$ is enriched, as depth increases; although for SiO$_2$ there is little change below about 250 m.

(c)   Dissolved silica is greatly enriched in pore waters below about 200–250 m, where the sediments contain a high proportion of biogenic siliceous material.

ACKNOWLEDGEMENTS

The author wishes to thank the following: Prof. Tim Jickells for helpful discussion and advice on Chapter 2; Prof. Paul Pearson for comment on Chapter 4; Dr Rachel Mills for her input to Chapter 5; and Angela Colling for advice and comment on the whole Volume. I would also like to thank John Wright for writing *Ocean Chemistry*, a precursor title in this Series and part of which has found its way into this new Volume.

The structure and content of the Series as a whole owes much to our experience of producing and presenting the first Open University course in Oceanography (S334) from 1976 to 1987. We are grateful to those people who prepared and maintained that Course, to the tutors and students who provided valuable feedback and advice and to Unesco for supporting its use overseas.

Every effort has been made to contact copyright holders. If any have been inadvertently overlooked, the publishers will be pleased to make the necessary arrangements at the first opportunity. Grateful acknowledgement is made to the following sources for permission to reproduce material within this book.

Figures 1.1, 1.4 Integrated Ocean Drilling Program; *Figures 1.2, 1.3* Institute of Oceanographic Sciences; *Figure 1.5a* Jeremy Young, Natural History Museum; *Figure 1.5b* NASA/Goddard Space Center/Orbimage/ SeaWiFS Project; *Figure 1.6a* Angela Hayes, PhD Thesis, 21 March 2000, University of Southampton; *Figure 1.6b* Matthew Olney, University College London; *Figure 1.7* Science Photo Library Ltd; *Figure 1.8a* copyright © Microscopy; *Figures 1.8b, 1.9* Wim van Egmond; *Figure 1.11* W. G. Deuser, Woods Hole Oceanographic Institution; *Figure 1.14* Paul Yates, Open University; *Figures 2.4, 2.6–2.8* Woods Hole Oceanographic Institution; *Figure 2.5* Dr. Richard Lampitt, Southampton Oceanography Centre; *Figure 2.11* Ridgwell, A. J. (2002) 'Dust in the Earth system', *Phil. Trans. Roy. Soc. Lond.*, **360**, 2905–2924, copyright © 2002 The Royal Society; *Figure 2.14* E. Suess (1988) *Nature*, **333**, Macmillan; *Figure 2.15* K. J. Orians and K. W. Bruland (1986) *Earth and Planetary Science Letters*, **78**, Elsevier; *Figures 2.20, 2.23, 2.24, 3.2* W. S. Broecker (1974) *Chemical Oceanography*, Harcourt Brace Jovanovich Inc.; *Figures 2.26, 3.6, 3.7* Steele, J. H. (ed.) (2001) *Encyclopedia of Ocean Sciences*, Academic Press; *Figure 2.30* Scottish Association for Marine Science (SAMS); *Figure 3.9* by permission of the British Geological Survey, © NERC, all rights reserved IPR/59-59C; *Figure 3.11* courtesy of C. D. Evans, British Geological Survey; *Figure 3.13d* Weaver, P. P. E. (2003) 'Northwest African continental margin', *Paleoceanography*, **18**(1) copyright © 2003 American Geophysical Union; *Figure 4.1* Photo: Robert Ginsburg; *Figure 4.2* Emiliani, C. (1955) 'Pleistocene temperatures', *Journal of Geology*, **63**, University of Chicago Press; *Figure 4.3* copyright © JAMSTEC; *Figure 4.4* Sclater, J. G. and McKenzie, D. P. (1973) *Geological Society of America Bulletin*, **84**(10), Geological Society of America; *Figure 4.6* Van Andel, T. H. (1994) *New Views on an Old Planet*, Cambridge University Press; *Figure 4.12* Chapman, M. R. *et al.* (1996) 'Faunal and alkenone reconstructions of subtropical North Atlantic surface hydrography....', *Paleoceanography*, **11**(3), American Geophysical Union; *Figure 4.13* Elderfield, H. and

INDEX